Springer Tracts in Advanced Robotics 127

Series editors

More information about this series at http://www.springer.com/series/5208

Daniel Sebastian Leidner

Cognitive Reasoning for Compliant Robot Manipulation

Springer

Daniel Sebastian Leidner
Institute of Robotics and Mechatronics
German Aerospace Center (DLR)
Wessling, Germany

ISSN 1610-7438 ISSN 1610-742X (electronic)
Springer Tracts in Advanced Robotics
ISBN 978-3-030-04857-0 ISBN 978-3-030-04858-7 (eBook)
https://doi.org/10.1007/978-3-030-04858-7

Library of Congress Control Number: 2018962373

This Springer imprint is published by the registered company Springer Nature Switzerland AG
The registered company address is: Gewerbestrasse 11, 6330 Cham, Switzerland

To my brother Björn

Foreword

Robotics is undergoing a major transformation in scope and dimension. From a largely dominant industrial focus, robotics is rapidly expanding into human environments and vigorously engaged in its new challenges. Interacting with, assisting, serving, and exploring with humans, the emerging robots will increasingly touch people and their lives.

Beyond its impact on physical robots, the body of knowledge robotics has produced is revealing a much wider range of applications reaching across diverse research areas and scientific disciplines, such as biomechanics, haptics, neurosciences, virtual simulation, animation, surgery, and sensor networks among others. In return, the challenges of the new emerging areas are proving an abundant source of stimulation and insights for the field of robotics. It is indeed at the intersection of disciplines that the most striking advances happen.

The *Springer Tracts in Advanced Robotics* (*STAR*) is devoted to bringing to the research community the latest advances in the robotics field on the basis of their significance and quality. Through a wide and timely dissemination of critical research developments in robotics, our objective with this series is to promote more exchanges and collaborations among the researchers in the community and contribute to further advancements in this rapidly growing field.

The monograph by Daniel Sebastian Leidner is based on the author's doctoral thesis. It conjugates the two research worlds of artificial intelligence and robotics in one original work on cognitive reasoning for compliant robot manipulation. Representation, planning, execution, and interpretation of constrained manipulation tasks naturally lead to the concept of intelligent physical compliance. These methodological issues are keenly discussed along with supporting practical issues, clearly demonstrating the potential of cognition-enabled service robots.

Rich of examples developed by means of extensive experimentation on a humanoid robotic platform both on ground and in a space context, this volume was the winner of the 2018 Georges Giralt PhD Award for the best doctoral thesis in Europe. A very fine addition to the STAR series!

Naples, Italy
October 2018

Bruno Siciliano
STAR Editor

Preface

This manuscript is the result of six years of research conducted between 2011 and 2017 at the Institute of Robotics and Mechatronics at the German Aerospace Center (DLR). The findings were elaborated in cooperation with the Institute for Artificial Intelligence at the University of Bremen. This exceptional alliance between two of the world's leading research facilities on artificial intelligence and robotics enabled me to realize this work.

Accordingly, I would first like to express my utmost gratitude to my supervisors, Prof. Michael Beetz and Prof. Alin Albu-Schäffer, for the opportunity to live out this relationship under their guidance. In addition, I would like to thank Christoph Borst for his trust in my qualities as a robotics researcher that allowed me to work in one of the most inspiring environments. My special thanks go to Florian Schmidt, who introduced me to the world of robotics and continually supported me during my research. Furthermore, I would like to thank Dr. Alexander Dietrich, which whom I collaborated in several publications leading to a better understanding of the interconnections between high-level reasoning and low-level control. Thanks to Georg Bartels, for the interesting and fruitful discussions during our research stays, which resulted in many valuable ideas. Thanks to my former students, Peter Birkenkampf and Wissam Bejjani, for their support.

My deep gratitude goes to the entire support team of Rollin' Justin for the maintenance of the robot, which are Thomas Gumpert, Werner Friedl, Martin Heumos, Robert Burger, Alexander Beyer, and Dr. Jörg Butterfass.

Special thanks go to the METERON SUPVIS Justin team, who helped to push the robot to the limits, resulting in an excellent experiment. In particular, I would like to thank Dr. Neal Y. Lii, Peter Birkenkampf, Ralph Bayer, and Benedikt Pleintinger.

Moreover, I would like to thank the people proofreading this book: Dr. Neal Y. Lii, Dr. Alexander Dietrich, Dr. Maxime Chalon, and Dr. Freek Stulp.

Thanks to my parents Birgit and Lothar, who supported me during my entire life, always believing in my strengths. Finally, I would like to thank my wife Hanna for her love. Without her patience and her support while raising our newborn son Leon Maximilian, I would have been unable to finish this work.

Munich, Germany
May 2017

Daniel Sebastian Leidner

Contents

List of Used Symbols and Abbreviations

The used symbols appear in equations or within the description of algorithms. Their meaning remains thereby consistent throughout the entire manuscript, if not otherwise stated. The listed pseudocode shall be considered independent of any concrete implementation, yet it is derived from Python syntax. Assignments of variables are thereby denoted by the left arrow symbol (e.g., $\gamma \leftarrow 0$). List attachments are denoted by the concatenation operator, where angle brackets encompass the element to be attached (e.g., $\boldsymbol{X}^\frown \langle \boldsymbol{x}_{\mathrm{end}} \rangle$).

In the following, the most frequent quantities and abbreviations of prominent importance are listed. They may appear with different subscripts and superscripts, where a dot donates a total derivative with respect to time t. Scalar quantities are printed as plain letters (e.g., α, $l_{\max}$, $\mathcal{O}$). Vectors and matrices are bold (e.g., $\dot{\boldsymbol{x}}, \boldsymbol{f}_{\mathrm{ext}}$, $\boldsymbol{P}$). The meaning of the symbols is not further specified at this point, but detailed in the respective chapters they are introduced.

Symbols

α	Action Template
γ	Tier value for a particular action
δ	Step size
μ	Mean value
σ	Standard deviation
i, j, k	Indices for numbering and iterations
n	Node element in a graph
r	Reachability index
t	Time
x, y	Particle coordinates
N, M	Numbers (e.g., number of robot joints)
$\mathcal{O}$	Object of interest
V	Spring potential

$\boldsymbol{\lambda}$	Discretized A* map
$\boldsymbol{\tau}$	Vector of joint torques
$\boldsymbol{\varphi}$	Region of interest
$\boldsymbol{e}$	List of effects
$\boldsymbol{f}$	Vector of (generalized) Cartesian forces
$\boldsymbol{p}$	List of preconditions
$\boldsymbol{q}$	Link-side joint configuration
$\boldsymbol{x}$	Vector of Cartesian coordinates
$\mathcal{A}$	List of Action Templates
$\boldsymbol{\Phi}$	List of low-level robot commands
$\boldsymbol{\Omega}$	List of elemental robot operations
$\mathcal{C}$	Controller parameter vector
$\mathbf{D}$	Positive definite damping matrix
$\mathcal{G}$	Geometric state of the environment
$\boldsymbol{J}$	Jacobian matrix
$\mathcal{P}$	Particle distribution state
$\mathcal{S}$	Symbolic state of the environment
$\mathcal{T}$	Transition vector (symbolic or geometric)
$\boldsymbol{Q}$	Configuration space trajectory
$\boldsymbol{X}$	Cartesian trajectory

Abbreviations

act	Actual
cmd	Command
des	Desired
dev	Deviation
e.g.	Exempli gratia (for example)
eef	End effector
ext	External
i.e.	Id est (that is)
max	Maximum
min	Minimum
w.r.t.	With respect to
AI	Artificial intelligence
DLR	Deutsches Zentrum für Luft- und Raumfahrt (German Aerospace Center)
DOF	Degree(s) of freedom
ESA	European Space Agency
ISS	International Space Station
KDE	Kernel density estimation
METERON	Multi-purpose End-To-End Robotic Operation Network
MST	Minimum spanning tree

OpenRAVE	Open Robotics Automation Virtual Environment
PDDL	Planning Domain Definition Language
ROI	Region of interest
RRT	Rapidly Exploring Random Trees
SDG	Semantic directed graph
TCP	Tool center point
TSP	Traveling salesperson

List of Figures

List of Tables

List of Listings

List of Algorithms

Abstract

Physically compliant contact is a major element for many tasks in everyday environments. A universal service robot that is utilized to collect leaves in a park, polish a workpiece, or clean solar panels requires the cognition and manipulation capabilities to facilitate such compliant interaction. Evolution equipped humans with advanced mental abilities to envision physical contact situations and their resulting outcome, dexterous motor skills to perform the actions accordingly, as well as a sense of quality to rate the outcome of the task. In order to achieve human-like performance, a robot must provide the necessary methods to represent, plan, execute, and interpret compliant manipulation tasks. This manuscript covers those four steps of reasoning in the concept of *intelligent physical compliance*.

The contributions advance the capabilities of service robots by combining artificial intelligence reasoning methods and control strategies for compliant manipulation. A classification of manipulation tasks is conducted to identify the central research questions of the addressed topic. Novel representations are derived to describe the properties of physical interaction. Special attention is given to wiping tasks which are predominant in everyday environments. It is investigated how symbolic task descriptions can be translated into meaningful robot commands. A particle distribution model is used to plan goal-oriented wiping actions and predict the quality according to the anticipated result. The planned tool motions are converted into the joint space of the humanoid robot Rollin' Justin to perform the tasks in the real world. In order to execute the motions in a physically compliant fashion, a hierarchical whole-body impedance controller is integrated into the framework. The controller is automatically parameterized with respect to the requirements of the particular task. Haptic feedback is utilized to infer contact and interpret the performance semantically. Finally, the robot is able to compensate for possible disturbances as it plans additional recovery motions while effectively closing the cognitive control loop. Among others, the developed concept is applied in an actual space robotics mission, in which an astronaut aboard the International Space Station (ISS) commands Rollin' Justin to maintain a Martian solar panel farm in a mock-up environment. This application demonstrates the far-reaching impact of the proposed approach and the associated opportunities that emerge with the availability of cognition-enabled service robots.

Chapter 1
Introduction

As technology continuously moves forward, universal service robots are envisaged to catch up with the human ability to master almost every task in everyday environments. Potential application domains range from personal assistance in domestic households to professional service in various areas, such as commercial estate, health-care applications, and public service. These robots will have to take care of the daily needs of the humans they are serving. They will have to perform everyday tasks such as cooking, organizing, and cleaning, but also support disabled people in the Activities of Daily Living (ADL) (Katz 1983), which are typically relatively simple tasks such as opening doors or giving assistance in drinking and eating. Furthermore, autonomous robots will revolutionize industrial manufacturing as predicted by Karel Čapek in the science fiction play *R.U.R. (Rossum's Universal Robots)* (Capek 1921). They will be deployed as intelligent co-workers to assist human workers in cumbersome tasks, and replace them in handling dangerous materials such as toxic waste. Similar to that, rescue robots may serve as an invaluable addition to disaster response scenarios as for example the nuclear incident of the Fukushima Daiichi Nuclear Power Plant (Nagatani et al. 2013). With the DARPA Robotics Challenge, it has already been showcased that a service robot may have been able to prevent a melt down (Pratt and Manzo 2013). Finally, intelligent space robot assistants will explore extraterrestrial environments, where tasks range from scientific studies, to the fully autonomous construction of planetary habitats for human scientists and possibly eventual long term human residents (Lii et al. 2015b). In conclusion, society agrees that

"There is a need for service robots."

However, until now the vision of universal service robots remains unfulfilled. In fact, robots are nowadays mainly deployed as ordinary machines to automate repetitive tasks in industrial manufacturing. These industrial robots have stiff structures to ensure high positioning accuracy and high speed. They cannot sensitively interact

D. S. Leidner, *Cognitive Reasoning for Compliant Robot Manipulation*, Springer Tracts in Advanced Robotics 127,
https://doi.org/10.1007/978-3-030-04858-7_1

Fig. 1.1 Artist's conception of a universal service robot cleaning a public area

with their environment as they lack compliant structures and the mandatory sensing capabilities to perceive the environment visually or measure contact forces during task execution. Consequently, this limits the achievable set of tasks to simple scenarios with deterministic outcome such as pick-and-place tasks. To this end, these robots can be considered a rigid assembly of actuators that operate without environmental feedback.

In contrast, most tasks in everyday environments cannot be accomplished with traditional industrial robots. For example, collecting leaves in a park as visualized in Fig. 1.1 demands compliant behavior to guide a rake along various surfaces such as hard cobblestone or soft meadows. Meanwhile, the robot has to constantly monitor the applied contact force and adapt to uncertainties by utilizing a feedback-driven control strategy. This requires compliant robotic manipulators, haptic perception, and carefully parametrized control strategies respectively. This signifies that

"There is a need for physical compliance."

In order to accomplish a chore such as collecting leaves properly, a robot needs advanced cognitive reasoning capabilities to understand the world and the objects it interacts with. The robot needs to know how to operate the tools to successfully accomplish the task. It has to perceive the leaves and (potentially dynamic) obstacles, reason about the desired world state (i.e. get the leaves accumulated), and plan goal-oriented tool motions in order to achieve the desired effects. Mastering a task like this requires a robot to solve variations of the problem in arbitrary environments. In particular, it has to be able to alter its behavior w.r.t. the current state of the world for every trial that is carried out. In order to truly understand the nature of the task, a robot must know *what* it is doing, *how* the actions have to be performed, and *why* it is executing them. Moreover, the robot needs to be able to analyze the task outcome and rate the performance. This way a robot will eventually be able to improve the result by questioning its own decisions and schedule additional actions. To achieve such cognitive abilities, it is necessary to embed the necessary reasoning methods deeply within the control programs of the robot by means of artificial intelligence (AI) reasoning methods. In other words,

"There is a need for artificial intelligence."

Even though robots have reached a high level of technical maturity in laboratory environments by now, they are still far from being deployable to any of the envisaged application domains. This is due to the fact that many tasks that are common and mundane for humans constitute a challenging problem for robots, both in terms of cognitive reasoning and motor skills. In particular, non-traditional robot tasks outside of factory buildings require compliant contact and show task outcome that is difficult to model. That includes the example of sweeping with a rake and other *wiping tasks*, as they occur in cleaning and manufacturing applications. While modern robots may be able to physically interact compliantly with their environment, they lack the cognitive abilities to interact as sophisticated with the world as humans do. In fact, real world applications demand compliant contact behavior *and* intelligent reasoning methods. This statement leads to the fundamental hypothesis:

"There is a need for intelligent physical compliance."

That is, a robot has to reason about compliant contact during every reasoning step as visualized in Fig. 1.2. In particular, this work combines cognitive reasoning methods from AI research and control mechanisms of robotics research to tackle the inherent issues. In particular, a robot has to

1. *Represent* the actions, motions and the resulting effects of compliant interaction.
2. *Plan* the task execution symbolically and geometrically w.r.t. the desired effect.
3. *Execute* the actions with the right parameterization of the compliance controller.
4. *Interpret* the task performance based on semantically enriched haptic information.

Unfortunately, researchers from AI and robotics have adopted different viewpoints on robotic manipulation. AI reasoning methods often consider planning issues from a *high-level*[1] cognitive science perspective. Manipulation tasks are typically represented from a scheduling point of view (Ghallab et al. 1998), where actions are modeled in a descriptive manner with detailed symbolic pre-conditions and effects. If an AI researcher is asked about the problem of collecting leaves in a park, the answer may sound as follows:

> First of all, the domain of the problem needs to be defined by means of an ontology. For example, one has to specify that a rake is a tool to collect leaves and it has to be carried by the robot to do so. The effect would be that the leaves are accumulated afterwards. Knowledge-based reasoning strategies have to be applied to infer that a rake has a certain probability to be picked up from a garden shed. A logic planner can schedule the actions accordingly.

[1] The term high-level refers to the abstract nature of manipulation tasks w.r.t. the overall task outcome.

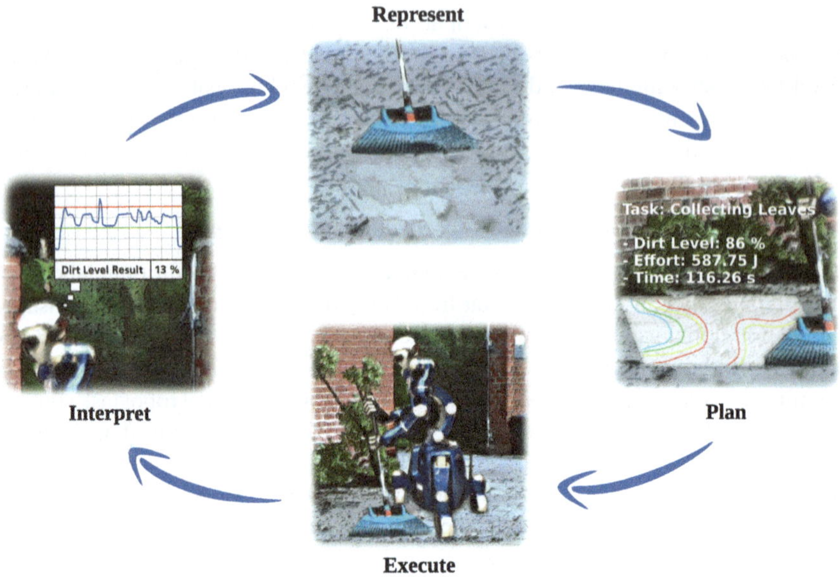

Fig. 1.2 The concept of Intelligent Physical Compliance

This AI perspective describes exactly *what* the requirements are, and what has to be done to achieve the desired effect, however, AI reasoning methods do not describe *how* to successfully execute the sweeping task. The concrete process model of the cleaning action is typically considered a black box that generates the desired outcome. The reasoning methods do not concern themselves with the physical procedure that the robot needs to firmly push the tool while applying a specific force profile onto the surface throughout the motion. Subsequently, AI reasoning systems are typically incapable of explaining and correcting errors caused by sub-optimal motion parameterization.

In robotics research, manipulation of objects is usually considered as a motion planning and control problem. For motion planning, the main constraint to be satisfied is to find collision-free paths to reach, pick up, and transfer the object to the destination. The *low-level*[2] robot controller is designed to execute the planned path as exactly as possible by means of high-rate control cycles. A control engineer may formulate the problem of collecting leaves in a park as follows:

> Sweeping motions can be realized by means of whole-body impedance control. The tool motion would thereby be described by a moving virtual equilibrium point for the tool center of the rake. A virtual potential force aligns the tool w.r.t. the curvature of the target surface. Defining an appropriate Cartesian stiffness at the end-effector results in compliant contact behavior that allows to cope with environmental uncertainties and external disturbances.

[2]The term low-level refers to specific properties describing the details of the manipulation procedure.

While this explanation describes *how* the task is executed, robotic control methods do not know *what* the abstract goal of the contact is, as they lack abstract task information. It is neither specified what a rake is nor what a proper outcome of the action should look like. The effects to the environment are mostly neglected in robotics research. Traditionally, robots are clueless about the purpose of their motions, and they are not aware of the resulting changes to the world.

Solving everyday manipulation tasks that require compliant contact such as the accumulation of leaves in a park is a multi-facet problem. It can be considered from many perspectives. The complexity of compliant manipulation tasks is analyzed in Sect. 1.1, in order to get a more profound understanding of the problem.

1.1 A Survey on Everyday Compliant Manipulation

Research on robotic manipulation was dominated by simple fetch-and-carry tasks for a long time. This domain was first introduced with the seminal blocks world benchmark developed for the Shakey system (Nilsson 1984). Since then, robotics research demonstrates scenarios that mostly include only very simple pick-and-place tasks. The involved actions are very simple in nature and result in discrete instantaneous effects. A successful task execution demands limited reasoning capabilities. A robot has to move toward the object until it is reachable, it has to know the shape of the object in order to plan a suitable grasp configuration for the hand or gripper, and it has to find a suitable spot on the target surface to place the object afterwards. Consequently, robotic agents which reason about this domain can abstract away from how actions are executed and model them using relatively simple representations. If anything, compliance is typically considered for safety reasons only.

In contrast, there are many more cases where compliant contact behavior is mandatory for a successful task execution. For example, in case of high impact forces as observed when hitting a nail with a hammer, humans actively adapt their joint stiffness by relaxing their muscles to maximize the range of motion upon impact to dissipate the impact energy (Grebenstein 2012). This behavior prevents damage of joints, bones, and muscles. Another type of deliberate compliance is observed in insertion tasks (Stemmer et al. 2007). For example, a person inserting a power plug will adapt the joint stiffness to automatically align the pose of the plug to the power outlet. A third example is given in wiping tasks, such as cleaning a mug with a sponge (Leidner et al. 2016b). In this example, lateral forces are applied to remove dirt and absorb water while moving the sponge along the surface of the mug. To do so, one has to act compliantly in order not to destroy the mug while adapting the sponge carefully to the target curvature. Robots performing these kind of tasks have to actively alter their stiffness parameterization w.r.t. individual task requirements, in order to get into deliberate physical contact with the environment.

Especially *wiping tasks* pose an interesting case of compliant manipulation. Based on the findings of a recently published study (Cakmak and Takayama 2013), wiping takes part in almost half of all daily household chores. The motivation for this study was to identify the main set of skills for future service robots. The authors have analyzed daily chore lists and found out that cleaning is one of the most frequent household chores. In particular, 49.8 % of the investigated chores were related to *wiping surfaces* of objects, furniture, or rooms. The Cambridge dictionary lists the verb *"to wipe"* as

> to slide something, especially a piece of cloth, over the surface of something else, in order to remove dirt, food, or liquid.[3]

According to this definition, the main actors in wiping tasks constitute of the *tool* (i.e. the cloth), the *surface* on a target object, and particles or liquids of different kind, from now on to be referred to as the *medium*. Cleaning occurs in various forms with many different goals including the cleaning of rooms, furniture and other objects by wiping with a floor-cleaning cloth, a feather duster, or a kitchen sponge as used for washing the dishes. However, this definition limits wiping to the act of dirt removal, as cleaning is the most prominent example wiping task in domestic environments. It can also be found in many of the earlier robotic application domains.

For this reason, vacuuming robots were among the first service robots to reach the consumer market to perform cleaning tasks in the household (Forlizzi and DiSalvo 2006). These robots are rather simple machines but are able to fulfill the tasks of vacuuming the floor (some of them even mopping). However, these robots are only able to execute the action they are designed for and therefore far from universal service robots.

In the long term, personal assistance robots may be used to clean immobilized persons which is considered as one of the most important ADLs. In this case, it is critical not to harm the person by carefully applying a sponge to the curvature of the skin yet not pushing too hard. In disaster areas, a robot may be commanded to remove debris by sweeping it aside in order to reach blocked regions easier. In planetary exploration robots have to solve various tasks which require compliance. One example is the maintenance of solar panels used to power a planetary research facility on Mars. As the harsh weather conditions on Mars regularly form sandstorms, it is necessary to free the sensitive solar panels from dust and sand by sweeping the dirt away.

While cleaning tasks represent a major branch of wiping actions, a more generic definition is desirable to cover the entire spectrum of actions that show characteristics of wiping. Besides cleaning, wiping may be considered as any task where a tool is slid along the surface of a target object, in order to manipulate arbitrary media. Based on this definition, the act of wiping represents a prototypical action for all domains that would benefit from robotic assistance. Cleaning-unrelated wiping actions are observed in many manufacturing applications including traditional manufacturing techniques such as woodworking, i.e. grinding a wooden plank, painting a wall, or applying mortar to a brick wall. Furthermore, compliant light-weight robots can be utilized to polish car parts in the automotive industry. The wide variation of wiping tasks is illustrated in Fig. 1.3.

[3] http://dictionary.cambridge.org/dictionary/english/wipe

Fig. 1.3 Examples of wiping actions observed in everyday environments

This point of view broadens the interpretation and understanding of wiping tasks toward a more generic definition. This makes wiping an even more interesting prototype of compliant manipulation. For this reason, the manuscript aims to identify suitable representations and reasoning strategies for wiping tasks. This investigations shall contribute to a more profound knowledge on compliance in robotic manipulation. Accordingly, this work first analyzcs wiping tasks to find out what makes compliant manipulation such a difficult, yet important problem. In the following, an attempt to redefine the act of wiping is presented:

> The act of wiping describes a compliant manipulation procedure, where a tool is slid along a target surface in contact by following task-oriented Cartesian workspace trajectories. During contact, task-specific forces are applied along the surface normal while a certain stiffness is adjusted to align with the surface curvature. A medium on the target surface is this way indirectly manipulated to satisfy the desired goal state.

Based on this definition, the manuscript shall investigate wiping tasks as the running example of compliant manipulation. The definition above is utilized as guideline to investigate wiping in a more generic way. A robot which is able to reason about a given task based on such abstract problem definitions shall be able to perform the majority of all tasks in everyday environments. However, this requires a robot to map high-level cognitive skills onto low-level dexterous manipulation abilities and back again. The following section highlights the issues that have to be incorporated to master the act of wiping.

1.2 Problem Statement

To fully master wiping tasks, a robot must understand every individual instance of the problem. It has to relate a new instance to previously executed ones and reason about similarities and differences for each trial. Investigating the entire spectrum of wiping tasks, it becomes apparent that many tasks matching the definition of wiping share the main characteristics (i.e. the sliding motion to manipulate a medium). However, most likely, they have different semantic goals. This makes it hard to develop generalized solutions matching every type of wiping task. In contrast to robots, humans can easily understand semantic differences and make crucial connections between related tasks. This knowledge originates from the hypothesis that humans represent internal models of their actions on a sub-symbolic[4] level, as suggested by Kawato (1999) in his research on neurobiology. A robot requires similar representations that can be accessed by means of semantically annotated labels and improved based on real-world experience.

A robot has to be able to plan generalized wiping motions with adaption to the particular environmental conditions. Humans are able to transfer the process model from one problem to another. For example, a human knows that collecting leaves with a rake or collecting shards of a broken mug with a broom is essentially a similar task. A human also knows that one cannot skim detergent from a window pane with a rake or a broom. Moreover, humans are aware of the effect of their actions. Consequently, humans plan their motion directly w.r.t. the desired effect, i.e. directly in the *effect-space*[5]. Similar cognitive reasoning mechanisms are required for robots. This way they would know exactly which motion has to be executed for a given problem instance.

In addition, it is crucial for a robot to be aware of the low-level control modalities of wiping tasks. That is, it has to know how much force has to be applied and how stiff it has to react in order to successfully accomplish the task. For example, only by applying the right amount of force and the correct stiffness, one is able to collect shards with a broom. If the applied force is too low, the bristles of the broom bend around the shards and the shards remain unmoved as it is visualized in Fig. 1.4. If too much force is applied, the bristles would bend too much and eventually relax abruptly such that the shards will scatter instead of accumulate. For this reason, a human would know that it is better to execute the task of sweeping with limited stiffness normal to the floor in order to adapt the broom motion automatically to a motion in parallel to the floor.

---

[4]Sub-symbolic parameters describe a process numerically, i.e. not only based on abstract symbols (words).

[5]The effect-space describes the manifold in which the effect occurs, e.g. the motion of dirt particles.

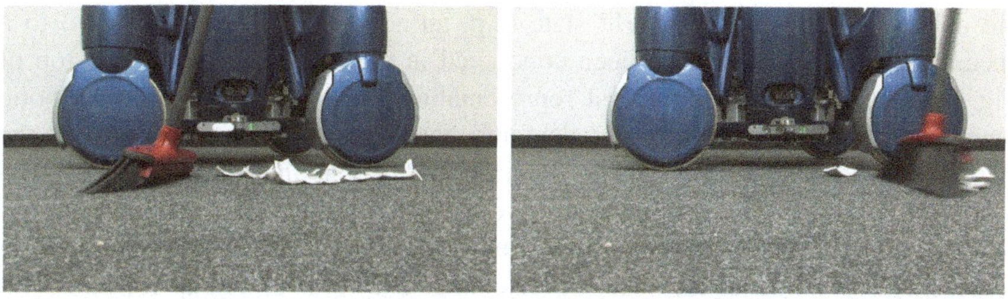

Fig. 1.4 Failed attempt to collect all shards of a broken mug. A robot has to be able to detect the effects of its actions and relate them to previously conducted predictions in order to estimate the task outcome qualitatively. This requires substantial reasoning on the contact situation, the applied contact force, and the compliance parameterization

Finally, a robot has to monitor and validate the task execution in order to ensure high quality. Humans are capable of interpreting the quality of wiping actions based on haptic perception, even with the absence of visual feedback. In particular, Flanagan et al. (2006) highlight that haptic feedback does not only provide humans with the information that contact occurred with the environment, but furthermore provides the basis for effect inference, performance ratings, and even the detection of performance errors and failure situations. In the following, it is outlined how this work investigates these issues by means of cognitive reasoning methods.

1.3 The Concept of Intelligent Physical Compliance

This manuscript aims to develop human-like, cognitive reasoning capabilities for robots in the context of wiping tasks, in order to eventually gain new insights into the development of a comprehensive set of cognition-enabled reasoning mechanisms for compliant manipulation tasks. The deeply integrated ability to represent, plan, execute, and interpret actions and effects of compliant manipulation is a key feature to master wiping tasks and other compliant manipulation task in general as it has already been introduced in the context of Fig. 1.2. In this work, this fundamental ability is called *Intelligent Physical Compliance*. The four elements of Intelligent Physical Compliance comprise the main research questions of this manuscript, outlined in the following sections.

1.3.1 Representation

The human ability to represent sub-symbolic knowledge (Kawato 1999) for a wide variety of physical tasks is one of the key features for many cognitive abilities, including planning and interpretation of interaction with the environment. Humans are able to relate a certain world state (e.g. an accumulated pile of leaves) to a specific action that must have been conducted in the past as visualized on top of Fig. 1.2. Accordingly, a suitable task representation is the key to reason about robotic manipulation in a generic way. The elements that need to be represented include the *objects* that participate in the task execution, the *actions* that these objects afford to a user or a robot, the *motions* to resolve these actions, and the *effects* that result from manipulating the object in the intended way, illustrated in Fig. 1.5. These representations influence all stages of reasoning. They are outlined as follows.

Actions: Actions relate motions to effects. An explicit representation of this relation enables a robot to reason about motions in a goal-oriented manner. That is, dedicated tool motions have to be planned in order to achieve the desired symbolic effects. Accordingly, this work shall develop a novel action representation that combines the symbolic description and the geometric procedure of actions. The so-called *Action Templates* formalize the symbol grounding-procedure for arbitrary robot skills. Action Templates provide the geometric procedure as executable Python code for a specific symbolic action definition written in PDDL. A symbolic planner is applied to generate a sequence of actions based on the symbolic header. The resulting sequence is afterwards revisited, where the executable body is parsed and populated with parameters according to the particular task instance.

Motions: In robotics, motions are typically represented by continuous, collision-free joint-space paths that enable a robot to relocate objects in the environment. However, collision-free motions are not applicable for compliant manipulation tasks such as collecting shards with a broken mug. In particular, wiping tasks demand contact with the environment. As such this work develops a representation of wiping motions that incorporates contact situations between a tool and a target surface in form of so called *Semantic Directed Graphs (SDG)*. In addition to the geometric motion, SDGs reincorporate the semantics of the problem. That is, the represented motions correlate with the desired goal state, e.g. the accumulation of leaves. The reasoning about motions is thereby tightly integrated with the representation of wiping effects.

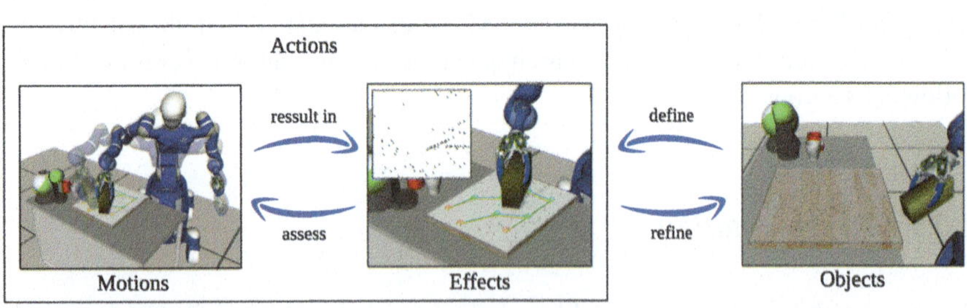

Fig. 1.5 Actions relate motions to effects. motions result in effects and in return, analyzing effects can be used to assess the performance of motions. Objects properties are used to parameterize actions and the outcome of an action can be used to refine object properties

Effects: Humans are able simulate their actions in advance through embodiment in order to envision the consequences of their motions (Svensson and Ziemke 2004). In return, humans are able to anticipate real-world effects based on these simulated models. While this reasoning seems natural for humans, simulation-based effect inference demands suitable representations for robots (Kunze 2014). In particular, a robot requires an effect model for every class of actions it is supposed to master. The representation developed for the example of wiping treats the medium as particles on a two-dimensional surface that react w.r.t. the desired semantic goal state upon contact with the tool. For example, while a sponge may push solid particles, liquids are absorbed by a sponge. The compliance parameters influencing the effect are therefore defined as object properties.

Objects: According to the concept of affordances introduced by Gibson (1986), the human demand to manipulate the environment is mainly driven by the objects that humans perceive. As a consequence, many research groups have tried to map the concept of affordance on robotic manipulation (Şahin et al. 2007; Stoytchev 2005; Hart et al. 2015; Cruz et al. 2015). This is a natural approach since effects to the environment are typically related to certain tool motions. Moreover, an object-centric interpretation of quality (e.g. symmetry, group formation, and coherency of objects) is a natural measure in everyday environments. For this reason, an object-centric approach shall be applied here to incorporate all requirements for a cognition-enabled planning, execution, and interpretation of compliant manipulation actions. With reference to the earlier definition (Sect. 1.1), the main objects that participate in the act of wiping are considered as the tool, the medium, and the target surface.

1.3.2 Planning

In order to master a task in arbitrary environments, a robot has to provide adaptive solutions. This involves symbolic and geometric planning techniques that are able to cope with the diversity of an entire task family. With reference to the example at hand, a robot has to provide generic planning solutions for arbitrary wiping tasks. A robot has to estimate the workload in advance, plan goal-oriented motions w.r.t. a specific geometric process model, and predict the task outcome respectively in order to execute the task most efficiently later on as seen on the right of Fig. 1.2. The robot has to perform geometric reasoning by means of task descriptions formulated in a purely semantic form (e.g. natural language). This problem formulation describes what is referred to as *hybrid planning*. This involves the processing of knowledge (Tenorth 2011), the semantic reasoning procedures on this information, and the symbol grounding problem (Harnad 1990), i.e. the mapping to geometric actions that result in the desired changes to the environment (Mösenlechner 2016).

From a symbolic planning perspective, the problem definition is initialized w.r.t. the current world state. Based on the state of the objects in the environment a symbolic

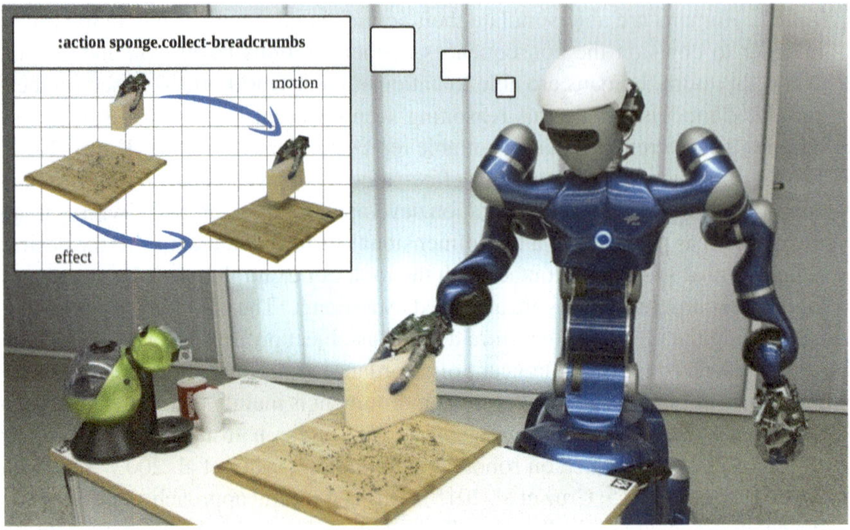

Fig. 1.6 The symbol grounding problem visualized at the example of collecting breadcrumbs by wiping along the surface with a sponge. The geometric action incorporates the wiping motion as well as the effect (i.e. the accumulated breadcrumbs). Both have to be associated with the semantic meaning

representation is generated. This representation serves as the basis for a symbolic planner such as the Fast Downward planner by Helmert (2006) to solve for a given high-level goal state. The resulting action sequence is the starting point to solve a task geometrically by means of motion planning procedures as they are provided by the motion planning framework OpenRAVE by Diankov (2010). However, in the context of wiping tasks this means that a robot has to deliberately plan contact situations instead of avoiding collisions which is usually the common sense in robotic motion planning.

Planning appropriate motions is not the only issue related to the symbol grounding problem. In addition, the effects of actions have to be predicted. However, this is mostly neglected in the literature. That is, a robot has to associate the effects of its action w.r.t. geometric changes to the environment as it is visualized in Fig. 1.6. The planned contact motions have to be correlated with the semantic goal state such that the desired effect eventuates. The approach introduced here is based on the particle representation for the medium. In particular, a set of Cartesian motion planning methods are associated with the semantically described goal states for the medium by means of constraint definitions. The particles are simulated such that they satisfy these constraints. The correlation between the final state of the particle distribution and the constraint definition is utilized to estimate the quality of the action. The outcome of this simulation does not only validate the feasibility of the robot motions, but also indicate the expected performance of the real-world execution.

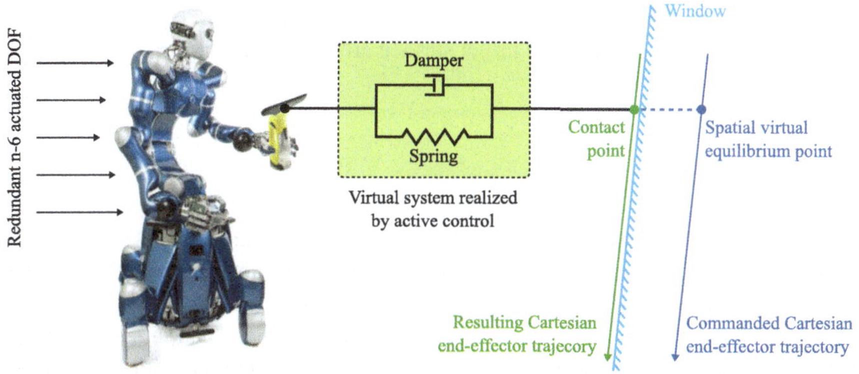

Fig. 1.7 Whole-body impedance control scheme based on Dietrich (2015). The redundant robot resolves the deviation of the spatial virtual equilibrium compliantly while the tool is in contact

1.3.3 Execution

One of the key aspect in automated robot manipulation is the question on how to parameterize actions to have their desired effects. At the example of wiping tasks this means that a robot has to parameterize the tool motions and the required compliance setting (e.g. force and stiffness) to manipulate a certain medium. The execution of robot motions (see bottom in Fig. 1.2) is typically considered as a control problem. Torque control strategies, such as impedance control (Hogan 1985) have to be applied for compliant motions where external contact with the environment is desired. The control strategy is outlined in Fig. 1.7 at the example of window cleaning. In a nutshell, the desired tool motion is provided as a Cartesian trajectory to the controller. The Cartesian controller realizes this motion while taking into account the specified stiffness and damping (Ott 2008). In addition, supplementary control tasks can be realized by exploiting the redundancy of the robot (Dietrich 2015). The resulting control torque adapts the tool orientation w.r.t. the curvature of the target surface as the tool gets in contact with the environment. This external disturbance results in the virtual Cartesian equilibrium point moving further "into" the object, such that the end-effector deviates from the commanded Cartesian trajectory and exerts a force. Moving in parallel to the target surface results in a wiping motion.

One major challenge with whole-body impedance controllers is the parameterization. First of all, several control objectives have to be organized in a hierarchy (e.g. Cartesian end-effector position, joint-space configuration, and self-collision avoidance). This hierarchy is task-dependent and has to be defined in advance. Second, the control parameters have to be designed in a way such that the desired effect occurs. That is, the reasoning framework has to parameterize the motions, the stiffness, and the maximum allowed force parameters based on the semantically described problem definition. In this work, the earlier introduced Action Templates are used to define the task hierarchy and the control parameters, respectively. The object-knowledge for the tool, the medium, and the target surface are thereby incorporated.

1.3.4 Interpretation

The final step toward Intelligent Physical Compliance is the semantic interpretation of physical contact, and the accompanied loop-closure of the cognitive control concept (Norman and Shallice 1980), which builds on the hypothesis that humans primarily take conscious action in tasks that require planning and decision making, and for error correction. This means that a robot would have to understand its motions, interpret the deviations introduced due to compliance, and assess the effect qualitatively to recover as sketched on the left in Fig. 1.2. The developed process models for wiping tasks highly exploit the fact that the impedance controller automatically aligns with the curvature of the target surface. While this exploitation prevents the development of complex process models that integrate dynamic simulations, the robot remains uncertain about the real contact until it touches the environment. To overcome this issue, the robot has to analyze telemetry data of previously executed actions. That is, it compares the desired contact with the actual contact situations. As soon as contact occurs, the external disturbance forces the robot motion to deviate according to the control law of the whole-body impedance controller. The tool aligns with the target surface and the redundant joints of the robot resolve the residual deviation of the planned trajectory. Consequently, the force at the end-effector increases until the parameterized level is reached. At this point, the robot is able to infer that the desired contact is established. The corresponding measurements are isolated in the logged data and utilized to estimate the contact, the position, and the orientation of the target surface, respectively.

The force information is furthermore utilized to evaluate the performance of the action, i.e. to assess the effect on the medium. Similarly to the planning step, the particle model is utilized for the evaluation. However, this time the contact force is available as additional source of information. The controller force is added to the particle simulation based on a likelihood function to simulate different variations of tool-medium-surface combinations. That is, measurements that show contact force values close to the desired force range have a higher likelihood to apply the effect to the particle distribution. Naturally, the robot has to distinguish desired contact from undesired contact, e.g. collisions with obstacles in the environment or human intervention as it is shown in Fig. 1.8. As the robot is now aware of the contact it is supposed to perform, the inverse method to the one that detects the target surface can be applied to detect undesired contact. The method allows to estimate the task performance not just as a scalar value, but it is able to estimate it spatially. This enables a robot to precisely plan additional repair motions based on the planning methods outlined earlier. As a result, the robot may question its own motions and react to execution errors appropriately.

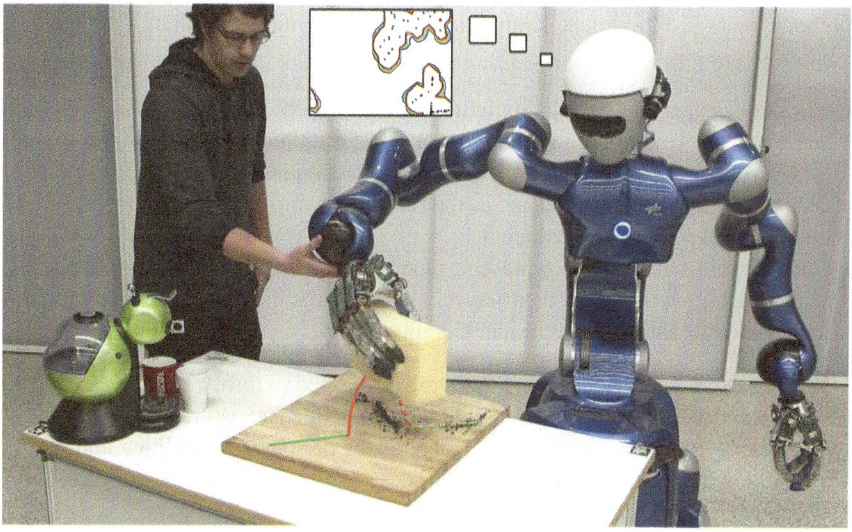

Fig. 1.8 The humanoid robot Rollin' Justin is disturbed during a wiping task

Finally, the qualitative values of the effect estimation are interpreted semantically. To do so, the reasoning methods are integrated into the openEASE framework (Beetz et al. 2015). The numerical values obtained during task execution (e.g. joint positions, force readings, object positions, and haptic percepts) are therefore post-processed and augmented with semantically meaningful labels. This procedure results in human-comprehensible information and can be inspected in form of narratives. For each time step, detailed information about spatial and temporal coherences is available w.r.t. objects, actions, and effects. Based on this, openEASE users can filter episodic memories by means of Prolog queries (Sterling and Shapiro 1994). A user may request the system to process relatively simple queries, such as "Visualize the trajectory segments in contact.", or ask even complex questions such as "Which sequence of motions result in a task performance greater than 85% ?". This last step closes the cognitive control loop from natural-language-like commands to human-comprehensible analysis.

1.4 Contributions

This manuscript combines cognitive reasoning techniques based on high-level abstraction, together with low-level manipulation capabilities of compliant service robots, to develop cognition-enabled reasoning methods for compliant manipulation. It contributes to the theoretic formulation of the problem and proves the proposed methods through experimental validation. The following list outlines the main contributions:

- An analysis of compliant manipulation tasks is presented to lay out the fundamentals for the development of cognition-enabled manipulation skills for compliant robots. Wiping tasks are identified as the most frequent compliant manipulation task. Therefore, wiping tasks act as the running example for compliant manipulation in the experiments conducted.
- The reasoning methods to represent, plan, execute and interpret compliant interaction constitute the major contribution. In summary, the developed methods form the concept of Intelligent Physical Compliance:

Representation: Action Templates are developed to form a suitable action representation to incorporate the semantic and geometric nature of robotic manipulation. This representation is independent of robot capabilities and can be utilized to describe even complex process models with context-sensitive intelligent behavior emerging from compliant manipulation tasks.
In order to investigate the effects of compliant manipulation, a particle model to simulate the outcome of wiping motions qualitatively is developed.

Planning: A framework is introduced to plan arbitrary action sequences represented by Action Templates. As part of this representation, process models are designed to generate contact motions that result in the desired effect of compliant manipulation tasks.
In the particular case of wiping actions, the introduced particle model is applied to reason about different planning methods summarized as Semantic Directed Graphs (SDG). Furthermore, the effect representation is utilized to predict the performance of the planned motions qualitatively by means of kinematics simulations.

Execution: The whole-body impedance control framework of the humanoid robot Rollin' Justin is combined with AI-based reasoning methods. In this regard, the control level can be parameterized in accordance to the needs of specific tasks. The control level shall be made aware of the desired situations and allow for deliberate compliant interaction.
In return, the control model provides relevant information in form of continuous sensor streams to allow for a meaningful semantic interpretation.

Interpretation: Reasoning methods to interpret episodic memories of compliant interaction semantically are introduced. Logged robot telemetry is analyzed w.r.t. the applied force and the position of a guided tool to isolate desired contact situations. By applying the effect model introduced earlier, this knowledge is used to validate the real-world outcome of the robot motions. The robot is similarly enabled to identify false positive contact states such as the collision with obstacles or human intervention.
The knowledge generated by the reasoning methods is annotated with semantically meaningful labels. This provides the possibility to query the acquired knowledge in a human-comprehensible form comparable to natural language,

i.e. Prolog queries. The data is made freely available to the openEASE community. This way, fellow researchers are able to query the conducted experiments and develop their own reasoning methods based on the findings of this manuscript.

- The concepts are validated on the humanoid robot Rollin' Justin. The main application domain throughout the manuscript is related to domestic household chores. However, as the robot is designed as a terrestrial and extraterrestrial service robot, this work describes an outlook on compliant manipulation tasks related to robotic planetary exploration within the METERON[6] SUPVIS Justin mission led by DLR with the partnering European Space Agency (ESA). During this mission, the humanoid robot Rollin' Justin is commanded from the International Space Station (ISS) to maintain a solar panel farm in a simulated Mars environment. This involves the regular cleaning of the panels. To achieve this, the entire reasoning framework is made accessible to an astronaut for human-robot interaction by means of an intuitive user interface in form of a tablet application. This user interface shares the knowledge-base of the robot with the astronaut while contemplating the current mission objectives.

1.5 Outline

This section recapitulates the content of the chapters and provides a reader's guide as it comments on the dependencies between them. In general, the manuscript is arranged w.r.t. the concept of Intelligent Physical Compliance such that the entire concept is introduced successively should the reader choose to read the document sequentially. The interested reader may, however, skip through some of the sections or revisit others to reconsider single aspects. An overview on the manuscript structure is given in Fig. 1.9. In addition, it lists the relevant publications in relation to the respective chapters.

Chapter 2 outlines the literature relevant to this manuscript. This includes an overview of the Rollin' Justin system, principles of motion planning and generation, symbolic planning and logic programming, as well as an outline on knowledge representation and reasoning. For a more detailed literature review w.r.t. the investigated reasoning methods, the reader may refer to the *end* of the individual chapters.

Chapter 3 recapitulates the problem of compliant manipulation and explains its importance for future service robots. It reviews the state of the art on taxonomies for robotic manipulation and makes an attempt to apply the most suitable ones to compliant manipulation. Eventually, the chapter develops the taxonomy for compliant manipulation tasks and inherits nine prototypical wiping tasks respectively.

[6] http://meteron.dlr.de/

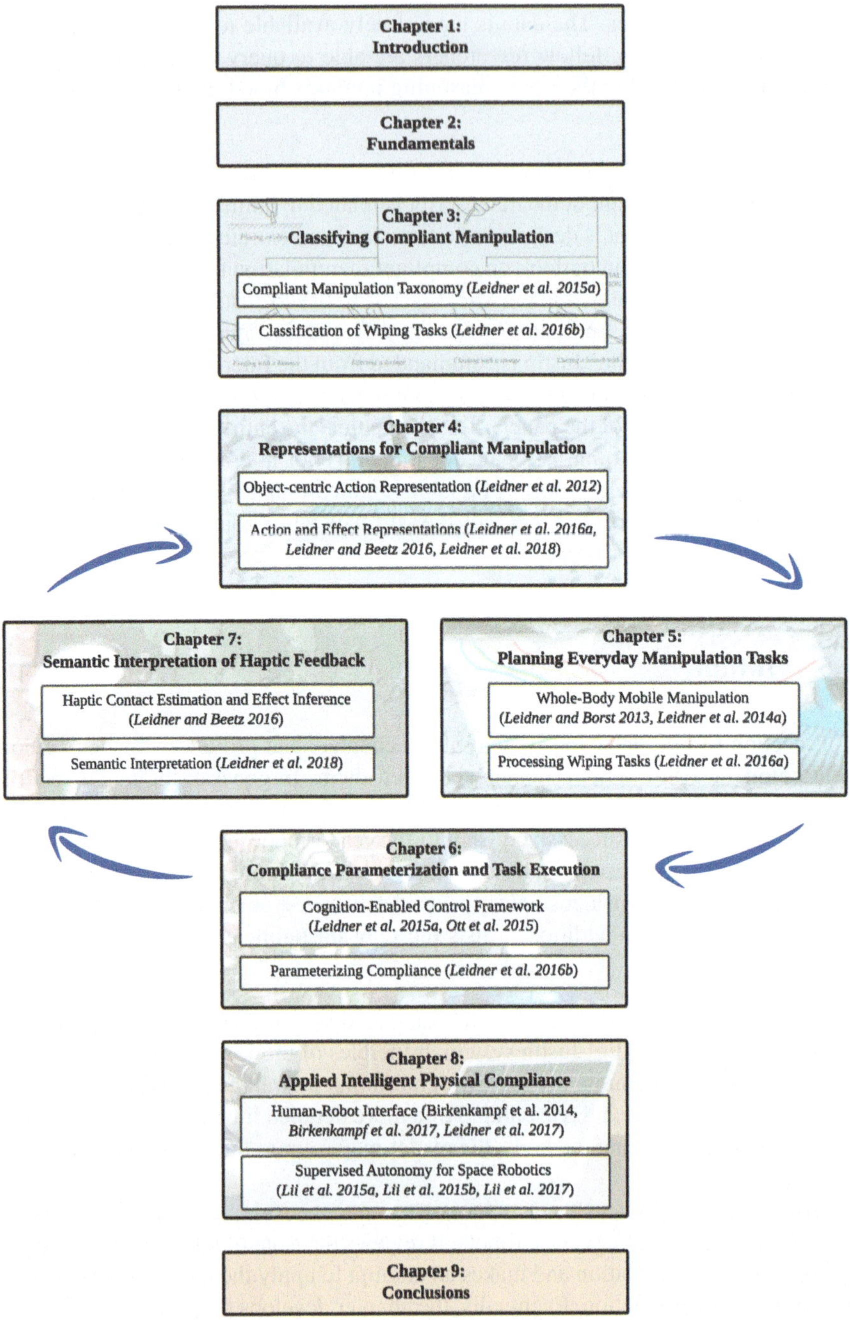

Fig. 1.9 Overview of the chapters. The associate research papers are listed accordingly. Chapters three to six are arranged w.r.t. the concept of Intelligent Physical Compliance

Chapter 4 describes the representations that are developed to plan, execute and interpret compliant manipulation tasks. This chapter introduces the fundamental representations on which all other chapters build on, namely the object-centric knowledge representation, the concept of Action Templates, the representation of wiping actions in form of SDGs, and the particle model to represent the effect of wiping motions.

Chapter 5 discusses the symbol grounding problem and a hybrid planning framework as potential solutions to it. It describes how the previously mentioned representations are utilized to reason about compliant manipulation symbolically and geometrically. The chapter concludes with a generic procedure to plan wiping motions with humanoid service robots and rate the task performance in the effect space. The reasoning methods are evaluated in several simulation experiments.

Chapter 6 elucidates the integration of the control framework into the AI-reasoning methods. It introduces the concept of impedance control and describes how the control level is parameterized and compliance is exploited to develop intelligent compliant Action Templates. The method is applied to three elaborate experiments executed by the humanoid robot Rollin' Justin.

Chapter 7 explains the methods to infer contact situations from logged telemetry streams and interpret them semantically. In particular, the representation of wiping effects is utilized to assess the task performance of previously executed contact motions qualitatively. This chapter closes the cognitive control loop as it describes how the newly acquired information about the task performance can be used to schedule additional actions to increase the task performance. The chapter then discusses the integration into the openEASE framework, which allows to infer nominal contact situations and human interventions based on semantically annotated episodic memories.

Chapter 8 outlines a use case for compliant manipulation tasks and introduces topics beyond the horizon of this manuscript, namely the development of cognition-enabled human-robot interfaces and their application to extraterrestrial space exploration. The METERON SUPVIS Justin mission is introduced as motivational example. This chapter may be of special interest for readers who are less familiar with robotics research or interested in real-world applications for intelligent compliant service robots. In particular, researchers from the field of planetary science may refer to this chapter for insights on robotic space exploration.

Chapter 9 concludes the manuscript with a summary of the contributions and discusses the benefits and limitations of the chosen approach. Furthermore, a discussion on the greater relevance of the developed methods is given w.r.t. the application to generic compliant manipulation. This chapter describes an overview of the manuscript and may therefore be read by readers who wish to start with a condensed version of the manuscript findings.

1.6 Publication Note

This manuscript was first released as dissertation at the University of Bremen (Leidner 2017). It builds on the findings of two journal articles, eleven conference papers published on major international congresses on robotics research and artificial intelligence, and one video submission, where some text passages are quoted verbatim. Part of this manuscript is currently under review as patent. These main publications are listed in Table 1.1. In addition, two workshop contributions have been published (Leidner et al. 2014a; Leidner and Borst 2013). Furthermore three conference papers (Roa et al. 2012; Leidner et al. 2015b; Hagengruber et al. 2017) have been co-authored, which are related but not addressed within this manuscript.

Table 1.1 Main publications related to this manuscript

Reference	Description
Journal: Leidner et al. (2016b)	Daniel Leidner, Alexander Dietrich, Michael Beetz, and Alin Albu-Schäffer. Knowledge enabled Parameterization of Whole-Body Control Strategies for Compliant Service Robots. *Autonomous Robots (AURO): Special Issue on Whole-Body Control of Contacts and Dynamics for Humanoid Robots*, 40(3): 519–536, 2016b
Journal: Ott et al. (2015)	Christian Ott, Alexander Dietrich, Daniel Leidner, Alexander Werner, Johannes Englsberger, Bernd Henze, Sebastian Wolf, Maxime Chalon, Werner Friedl, Alexander Beyer, et al. From Torque-ontrolled to Intrinsically Compliant Humanoid Robots. *Mechanical Engineering*, 3(2):7–11, 2015
Conference: Leidner et al. (2017)	Daniel Leidner, Neal Y. Lii, and Peter Birkenkampf. Context-aware Mission Control for Astronaut-Robot Collaboration. In *Proc. of the 14th Symposium on Advanced Space Technologies for Robotics and Automation (ASTRA)*, 2017
Conference: Birkenkampf et al. (2017)	Peter Birkenkampf, Daniel Leidner, and Neal Y. Lii. Ubiquitous User Interface Design for Space Robotic Operation. In *Proc. of the 14th Symposium on Advanced Space Technologies for Robotics and Automation (ASTRA)*, 2017
Conference: Lii et al. (2017)	Neal Y. Lii, Daniel Leidner, Peter Birkenkampf, Benedikt Pleintinger, Ralph Bayer, and Thomas Krueger. Toward Scalable Intuitive Teleoperation of Robots for Space Deployment with the METERON SUPVIS Justin Experiment. In *Proc. of the 14th Symposium on Advanced Space Technologies for Robotics and Automation (ASTRA)*, 2017
Conference: Leidner and Beetz (2016)	Daniel Leidner and Michael Beetz. Inferring the E ects of Wiping Motions based on Haptic Perception. In *Proc. of the IEEE/RAS International Conference on Humanoid Robots (ICHR)*, pages 461–468, 2016
Conference: Leidner et al. (2016a)	Daniel Leidner, Wissam Bejjani, Alin Albu-Schäff er, and Michael Beetz. Robotic Agents Representing, Reasoning, and Executing Wiping Tasks for Daily Household Chores. In *Proc. of the International Conference on Autonomous Agents and Multiagent Systems (AAMAS)*, pages 1006–1014, 2016a
Conference: Leidner et al. (2015a)	Daniel Leidner, Christoph Borst, Alexander Dietrich, and Alin Albu-Schäff er. Classifying Compliant Manipulation Tasks for Automated Planning in Robotics. In *in Proc. of the IEEE/RSJ International Conference on Intelligent Robots and Systems (IROS)*, pages 1769–1776, 2015a
Conference: Lii et al. (2015b)	Neal Y Lii, Daniel Leidner, André Schiele, Peter Birkenkampf, Benedikt Pleintinger, and Ralph Bayer. Command Robots from Orbit with Supervised Autonomy: An introduction to the METERON SUPVIS Justin Telerobotic Experiment. In *Proc. of the ACM/IEEE International Conference on Human-Robots Interaction (HRI)*, pages 53–54, 2015b

(continued)

Table 1.1 (continued)

Reference	Description
Conference: Lii et al. (2015a)	Neal Y Lii, Daniel Leidner, André Schiele, Peter Birkenkampf, Ralph Bayer, Benedikt Pleintinger, Andreas Meissner, and Andreas Balzer. Simulating an Extraterrestrial Environment for Robotic Space Exploration: The METERON SUPVIS Justin Telerobotic Experiment and the SOLEX Proving Ground. In *Proc. of the 13th Symposium on Advanced Space Technologies in Robotics and Automation (ASTRA)*, 2015a
Conference: Birkenkampf et al. (2014)	Peter Birkenkampf, Daniel Leidner, and Christoph Borst. A Knowledge-Driven Shared Autonomy Human-Robot Interface for Tablet Computers. In *Proc. of the IEEE/RAS International Conference on Humanoid Robots (ICHR)*, pages 152–159, 2014
Conference: Leidner et al. (2014b)	Daniel Leidner, Alexander Dietrich, Florian Schmidt, Christoph Borst, and Alin Albu-Schäffer. Object-Centered Hybrid Reasoning for Whole-Body Mobile Manipulation. In *Proc. of the IEEE International Conference on Robotics and Automation (ICRA), pages* 1828–1835, 2014b
Conference: Leidner et al. (2012)	Daniel Leidner, Christoph Borst, and Gerd Hirzinger. Things Are Made for What They Are: Solving Manipulation Tasks by Using Functional Object Classes. In *Proc. of the IEEE/RAS International Conference on Humanoid Robots (ICHR)*, pages 429–435, 2012
Video: Leidner and Dietrich (2015)	Daniel Leidner and Alexander Dietrich. Towards Intelligent Compliant Service Robots. In *Twenty-Ninth AAAI Conference on Articial Intelligence, AAAI Video Competition, 2015*. URL http://youtu.be/jglwgcz8iaM
Patent: Leidner and Birkenkampf (2017)	Daniel Leidner and Peter Birkenkampf. Verfahren zum Steuern eines Roboters, German patent application no. 10 2017 209 032.4, filed on May 30, 2017, 2017

References

Beetz, Michael, Moritz Tenorth, and Jan Winkler. 2015. Open-EASE—A Knowledge Processing Service for Robots And Robotics/AI Researchers. In *IEEE International Conference on Robotics and Automation (ICRA)*, pp. 1983–1990.

Birkenkampf, Peter, Daniel Leidner, and Neal Y. Lii. 2017. Ubiquitous User Interface Design for Space Robotic Operation. In *Proceedings of the 14th Symposium on Advanced Space Technologies for Robotics and Automation (ASTRA)*.

Birkenkampf, Peter, Daniel Leidner, and Christoph Borst. 2014. A Knowledge-driven Shared Autonomy Human-robot Interface for Tablet Computers. In *Proceedings of the IEEE/RAS International Conference On Humanoid Robots (ICHR)*, pp. 152–159.

Capek, Karel. 1921. *R.U.R. (Rossum's Universal Robots)*.

Cakmak, Maya, and Leila Takayama. 2013. Towards a Comprehensive Chore List for Domestic Robots. In *Proceedings of the ACM/IEEE International Conference on Human-robot Interaction (HRI)*, pp. 93–94.

Cruz, Francisco, German I Parisi, and Stefan Wermter. 2015. Contextual Affordances for Action-effect Prediction in a Robotic-cleaning Task. In *Proceedings of the Workshop on Learning Object Affordances at the IEEE/RSJ International Conference on Intelligent Robots and Systems (IROS)*.

Diankov, Rosen. 2010. *Automated Construction of Robotic Manipulation Programs*. PhD thesis, Carnegie Mellon University, Robotics Institute.

Dietrich, Alexander. 2015. *Whole-Body Impedance Control of Wheeled Humanoid Robots*. PhD thesis, Technische Universität München (TUM).

Forlizzi, Jodi and Carl DiSalvo. 2006. Service Robots in the Domestic Environment: A Study of the Roomba Vacuum in the Home. In *Proceedings of the 1st ACM SIGCHI/SIGART Conference on Human-robot Interaction (HRI)*, pp. 258–265.

Flanagan, J.Randall, Miles C. Bowman, and Roland S. Johansson. 2006. Control Strategies in Object Manipulation Tasks. *Current Opinion in Neurobiology* 16 (6): 650–659.

Ghallab, Malik, Adele Howe, Dave Christianson, Drew McDermott, Ashwin Ram, Manuela Veloso, Daniel Weld, and David Wilkins. 1998. PDDL—The Planning Domain Definition Language. *AIPS98 Planning Committee* 78 (4): 1–27.

Grebenstein, Markus. 2012. *Approaching Human Performance*. PhD thesis, Springer.

Gibson, James J. 1986. *The Ecological Approach to Visual Perception*. Routledge.

Hart, Stephen, Paul Dinh, and Kimberly Hambuchen. 2015. The Affordance template ROS Package for Robot Task Programming. In *Proceedings of the IEEE International Conference on Robotics And Automation (ICRA)*, pp. 6227–6234.

Harnad, Stevan. 1990. The Symbol Grounding Problem. *Physica D: Nonlinear Phenomena* 42 (1–3): 335–346.

Helmert, Malte. 2006. The Fast Downward Planning System. *Journal of Artifcial Intelligence Research* 26: 191–246.

Hagengruber, Annette Daniel Leidner, and Jörn Vogel. 2017. EDAN—EMG-Controlled Daily Assistant. In *Proceedings of the ACM/IEEE International Conference on Human-robot Interaction (HRI)*, pp. 409.

Hogan, Neville. 1985. Impedance Control: An Approach to Manipulation: Part I-Theory, Part II–Implementation, Part III–Applications. *Journal of Dynamic Systems, Measurement, and Control* 107: 1–24.

Katz, Sidney. 1983. Assessing Self-maintenance: Activities of Daily Living, Mobility, and Instrumental Activities of Daily Living. *Journal of the American Geriatrics Society* 31 (12): 721–727.

Kawato, Mitsuo. 1999. Internal Models for Motor Control and Trajectory Planning. *Current Opinion in Neurobiology* 9 (6): 718–727.

Kunze, Lars. 2014. *Naive Physics and Commonsense Reasoning for Everyday Robot Manipulation*. PhD thesis, Technische Universität München (TUM).

Lii, Neal Y, Daniel Leidner, André Schiele, Peter Birkenkampf, Benedikt Pleintinger, and Ralph Bayer. 2015b. Command Robots From Orbit with Supervised Autonomy: An Introduction to the METERON SUPVIS Justin Telerobotic Experiment. In *Proceedings of the ACM/IEEE International Conference on Human-robots Interaction (HRI)*, pp. 53–54.

Leidner, Daniel, Alexander Dietrich, Michael Beetz, and Alin Albu-Schäffer. 2016b. Knowledge-Enabled Parameterization of Whole-body Control Strategies for Compliant Service Robots. *Autonomous Robots (AURO): Special Issue on Whole-Body Control of Contacts and Dynamics for Humanoid Robots* 40 (3): 519–536.

Leidner, Daniel, Neal Y. Lii, and Peter Birkenkampf. 2017. Context-Aware Mission Control for Astronaut-robot Collaboration. In *Proceedings of the 14th Symposium on Advanced Space Technologies For Robotics And Automation (ASTRA)*.

Leidner, Daniel and Michael Beetz. 2016. Inferring the Effects of Wiping Motions based on Haptic Perception. In *Proceedings of the IEEE/RAS International Conference on Humanoid Robots (ICHR)*, pp. 461–468.

Leidner, Daniel, Christoph Borst, Alexander Dietrich, and Alin Albu-Schäffer. 2015a. Classifying Compliant Manipulation Tasks for Automated Planning in Robotics. In *Proceedings of the IEEE/RSJ International Conference on Intelligent Robots and Systems (IROS)*, pp. 1769–1776.

Leidner, Daniel, Alexander Dietrich, Florian Schmidt, Christoph Borst, and Alin Albu-Schäffer. 2014b. Object-Centered Hybrid Reasoning for Whole-body Mobile Manipulation. In *Proceedings of the IEEE International Conference on Robotics and Automation (ICRA)*, pp. 1828–1835.

Leidner, Daniel and Alexander Dietrich. 2015. Towards Intelligent Compliant Service Robots. In *Twenty-Ninth AAAI Conference on Artificial Intelligence, AAAI Video Competition*. http://youtu.be/jgIwgcz8iaM.

Leidner, Daniel Sebastian. 2017. *Cognitive Reasoning for Compliant Robot Manipulation*. PhD thesis, University of Bremen.

Leidner, Daniel and Christoph Borst. 2013. Hybrid Reasoning for Mobile Manipulation based on Object Knowledge. In *Proceedings of the Workshop on AI-based Robotics at IEEE/RSJ International Conference on Intelligent Robots and Systems (IROS)*.

Leidner, Daniel, Selma Music, and Armin Wedler. 2015b. Robotic Deployment of Extraterrestrial Seismic Networks. In *Proceedings of the 13th Symposium on Advanced Space Technologies in Robotics and Automation (ASTRA)*.

Lii, Neal Y., Daniel Leidner, Peter Birkenkampf, Benedikt Pleintinger, Ralph Bayer, and Thomas Krueger. 2017. Toward Scalable Intuitive Teleoperation of Robots for Space Deployment with the METERON SUPVIS Justin Experiment. In *Proceedings of the 14th Symposium on Advanced Space Technologies for Robotics and Automation (ASTRA)*.

Leidner, Daniel, Wissam Bejjani, Alin Albu-Schäffer, and Michael Beetz. 2016a. Robotic Agents Representing, Reasoning, and Executing Wiping Tasks for Daily Household Chores. In *Proceedings of the International Conference on Autonomous Agents and Multiagent Systems (AAMAS)*, pp. 1006–1014.

Leidner, Daniel, Christoph Borst, and Gerd Hirzinger. 2012. Things are Made for What they are: Solving Manipulation Tasks by Using Functional Object Classes. In *Proceedings of the IEEE/RAS International Conference on Humanoid Robots (ICHR)*, pp. 429–435.

Leidner, Daniel, Peter Birkenkampf, Neal Y. Lii, and Christoph Borst. 2014b. Enhancing Supervised Autonomy for Extraterrestrial Applications by Sharing Knowledge Between Humans and Robots. In *Proceedings of the Workshop on How to Make Best Use of a Human Supervisor for Semi-autonomous Humanoid Operation at IEEE-RAS International Conference on Humanoid Robots (ICHR)*.

Lii, Neal Y, Daniel Leidner, André Schiele, Peter Birkenkampf, Ralph Bayer, Benedikt Pleintinger, Andreas Meissner, and Andreas Balzer. 2015a. Simulating an Extraterrestrial Environment for Robotic Space Exploration: The METERON SUPVIS Justin Telerobotic Experiment and the SOLEX Proving Ground. In *Proceedings of the 13th Symposium on Advanced Space Technologies in Robotics and Automation (ASTRA)*.

Leidner, Daniel and Peter Birkenkampf. 2017. Verfahren zum Steuern eines Roboters, German patent application no. 10 2017 209 032.4, filed on May 30, 2017.

Mösenlechner, Lorenz. 2016. *The Cognitive Robot Abstract Machine*. PhD thesis, Universität München.

Çakmak, Maya, Erol Şahin, Mehmet R. Doğar, Emre Uğur, and Göktürk Üçoluk. 2007. To Afford or Not to afford: A New Formalization of Affordances Toward Affordance-based Robot Control. *Adaptive Behavior* 15 (4): 447–472.

Nilsson, Nils J. 1984. *Shakey the Robot*. DTIC Document: Technical report.

Norman, Donald A., and Tim Shallice. 1980. *Attention to Action: Willed and Automatic Control of Behavior*. DTIC Document: Technical report.

Nagatani, Keiji, Seiga Kiribayashi, Yoshito Okada, Kazuki Otake, Kazuya Yoshida, Satoshi Tadokoro, Takeshi Nishimura, Tomoaki Yoshida, Eiji Koyanagi, Mineo Fukushima, et al. 2013. Emergency Response to the Nuclear Accident at the Fukushima Daiichi Nuclear Power Plants Using Mobile Rescue Robots. *Journal of Field Robotics* 30 (1): 44–63.

Ott, Christian. 2008. *Cartesian Impedance Control of Redundant and Flexible-joint Robots*, vol. 49. Springer Tracts in Advanced Robotics Berlin Heidelberg: Springer Publishing Company.

Ott, Christian, Alexander Dietrich, Daniel Leidner, Alexander Werner, Johannes Englsberger, Bernd Henze, Sebastian Wolf, Maxime Chalon, Werner Friedl, Alexander Beyer, et al. 2015. From Torque-ontrolled to Intrinsically Compliant Humanoid Robots. *Mechanical Engineering* 3 (2): 7–11.

Pratt, Gill, and Justin Manzo. 2013. The DARPA Robotics Challenge. *Robotics & Automation Magazine, IEEE* 20 (2): 10–12.

Roa, Maximo A, Max J Argus, Daniel Leidner, Christoph Borst, and Gerd Hirzinger. 2012. Power Grasp Planning for Anthropomorphic Robot Hands. In *IEEE International Conference on Robotics and Automation (ICRA)*, pp. 563–569.

Stemmer, Andreas, Alin Albu-Schäffer, and Gerd Hirzinger. An Analytical Method for the Planning of Robust Assembly Tasks of Complex Shaped Planar Parts. In *Proceedings of the IEEE International Conference on Robotics And Automation (ICRA)*, pp. 317–323.

Stoytchev, Alexander. 2005. Behavior-Grounded Representation of Tool Affordances. In *Proceedings of the IEEE International Conference on Robotics and Automation (ICRA)*, pp. 3060–3065.

Sterling, Leon and Ehud Y Shapiro. 1994. *The art of Prolog: Advanced Programming Techniques*. MIT press.

Svensson, Henrik, and Tom Ziemke. 2004. Making Sense of Embodiment: Simulation Theories and the Sharing of Neural Circuitry Between Sensorimotor and Cognitive Processes. In *Proceedings of the 26th Annual Meeting of the Cognitive Science Society*, pp. 1309–1314.

Tenorth, Moritz M. 2011. *Knowledge Processing for Autonomous Robots*. PhD thesis, Universität München.

Chapter 2
Fundamentals

The successful performance of diverse wiping tasks requires dexterous and compliant robots, appropriate motion planning and generation skills to establish desired contact, methods for symbolic planning and logic programming, as well as representations and mechanisms to reason about the components and effects of the actions. These domains form the prerequisites which this work builds on. As such, they are outlined in the following sections. A general overview on compliant manipulation in AI and robotics is presented in Sect. 2.1. The humanoid robot Rollin' Justin is introduced in Sect. 2.2. The available algorithms for motion planning and generation are investigated in Sect. 2.3. Symbolic planning and the related field of logic programming is revisited in Sect. 2.4. Finally, knowledge representations and the accompanied reasoning mechanisms are reviewed in Sect. 2.5. A literature review for the concrete implementation of the systems presented in this document can be found *at the end* of each respective chapter.

2.1 Physical Compliance in AI and Robotics

In the last decades, there has been remarkable progress in the mechanical design and control of compliant robotic manipulators. The mechatronic design of torque controlled light-weight manipulators such as the light-weight robot III (LWR III) (Hirzinger et al. 2002), developed at the *German Aerospace Center (DLR), Institute of Robotics and Mechatronics*, enable robots to compliantly interact with their environment. The advanced robotic systems that emerged from this development can truly be considered as the prototypes of universal service robots. Especially humanoid and anthropomorphic robots are predestined to support humans in everyday environments as they are designed to mimic human capabilities. Groundbreaking examples of compliant humanoid robots are *Twendy-One* developed at Waseda University (Iwata and

D. S. Leidner, *Cognitive Reasoning for Compliant Robot Manipulation*, Springer Tracts in Advanced Robotics 127,
https://doi.org/10.1007/978-3-030-04858-7_2

Sugano 2009), the latest version of *Asimo* by Honda Research (Sakagami et al. 2002), NASA's *Robonaut 2* (Diftler et al. 2011), the *Baxter* robot distributed by Rethink Robotics (Fitzgerald 2013), and the humanoid robots *Toro* (Englsberger et al. 2014) and *Rollin' Justin* (Borst et al. 2009) developed at DLR. Based on these systems, there has also been significant progress in the field of compliant control of complex robotic systems. Impedance control strategies both in joint space and Cartesian space allow for smooth motions and soft contact (Albu-Schäffer et al. 2007; Ott 2008). Advanced whole-body impedance control frameworks (Dietrich et al. 2012) facilitate complex mobile manipulation procedures while simultaneously satisfying passivity, stability, and safety of the system at all times. This development enables these robots to closely co-operate with humans and get in soft contact with the physical world which is one of the key features for everyday robotic manipulation.

On the other hand, AI plays a major role in the advancement of robotic systems. For this reason, AI research has focused on the field of robotics early on. One of the first robots with cognitive reasoning capabilities was the *Shakey* system (Nilsson 1984). The program of Shakey included automated planning, navigation, and error recovery strategies. The system could manipulate blocks fully autonomously by commanding simple actions, e. g. "`(push box1, (14.1, 22.7))`", or specifying abstract high level goals in form of written sentences such as "`use box1 to block door4 from room2`". In order to adapt robot actions w. r. t. previously unseen environments, there has been significant research efforts in the field of machine learning. As a result laboratory robots are able to enhance known actions by applying reinforcement leaning strategies (Stulp et al. 2012) and even learn new actions from human demonstration (Billard et al. 2008). To manage the vast information arising from robotic manipulation, AI research has recently shown strong interest in knowledge-based reasoning systems such as the KnowRob framework (Tenorth and Beetz 2009). This framework is able to provide a robot with the necessary information to represent, plan and execute everyday manipulation tasks. In addition, the knowledge processing service openEASE (Beetz et al. 2015) provides a platform to annotate episodic memories of robotic manipulation with semantic labels. To this end, AI and robotics researchers are able to query abstract questions in order to investigate big data manipulation sequences in a human comprehensible form. Altogether these developments aim to replicate human-like reasoning capabilities to eventually create *cognition-enabled* robots that are able to catch up with human intelligence. However, the applications that are typically investigated by the research groups in AI and robotics are mostly composed of rather simple fetch-and-carry tasks.

Compared to fetch-and-carry tasks, compliant manipulation tasks demand significantly richer and more complex actions. Detailed knowledge about the task and the involved objects is mandatory. A robot has to reason symbolically and geometrically about the parameterization in advance w. r. t. the desired goal state and the current state of the environment. For example, depending on how a wiping motion is performed in terms of direction of motion, tool alignment, and applied force one can remove sticky dirt, dry a wet surface, or paint a wall. Consequently, the same motion can produce widely varying effects depending on how it is executed. In order to accomplish the respective tasks successfully a robot must carefully select the motion

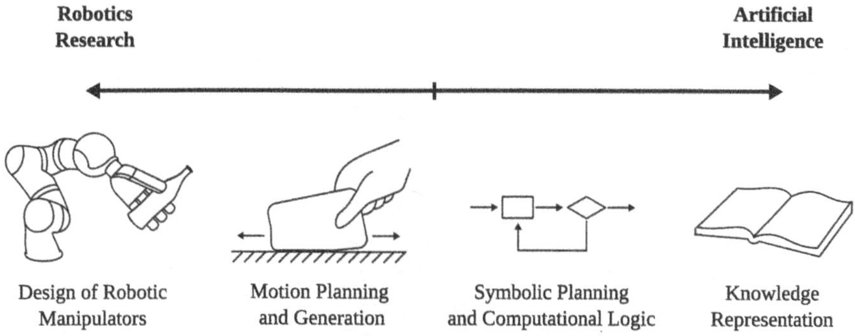

Fig. 2.1 Excerpt of the research topics of AI and robotics

and force parameters in a continuous parameter space within the bounds of the task. Moreover, the successful execution of wiping tasks requires a substantial amount of geometric reasoning as the purpose of an action might be the collection of particles, the spreading of particles, or the elimination of particles. This reasoning also applies to the verification procedure, as the effects are non-trivial changes to the manipulated objects. To overcome the complexity of compliant manipulation tasks, the research efforts of AI and robotics should be combined. However, the two research fields cover completely different perspectives on the problem definition. As a consequence, the research directions deviate as illustrated in Fig. 2.1. It sketches an excerpt of four core research topics: The design and construction of compliant robotic manipulators, planning and generation of robot motions, logic programming and symbolic planning, and finally, the topic of knowledge representation and reasoning.

2.2 The Humanoid Service Robot Rollin' Justin

Compliant manipulation tasks require robots with above average reachability, in order to incorporate uncertainties introduced by the contact with the environment. Especially wiping motions require high manipulability in order to achieve the desired effects for large areas. Accordingly, humanoid robots are predestined for this kind of tasks. For this reason, the humanoid robot Rollin' Justin[1] is utilized in this work to validate the developed concepts.

This section introduces the architecture and the capabilities of Rollin' Justin (Ott et al. 2006; Borst et al. 2007, 2009), which serves as the main concept realization and validation platform in this work. The overall structure of the robot resembles an anthropomorphic upper body mounted on a wheeled mobile base as it is shown in Fig. 2.2. With a height of about 1.9 m it roughly matches the dimensions of a male

[1]The upper body of the Rollin' Justin was finished *just in* time for its first presentation at the AUTOMATICA trade fair 2006 in Munich (Ott et al. 2006).

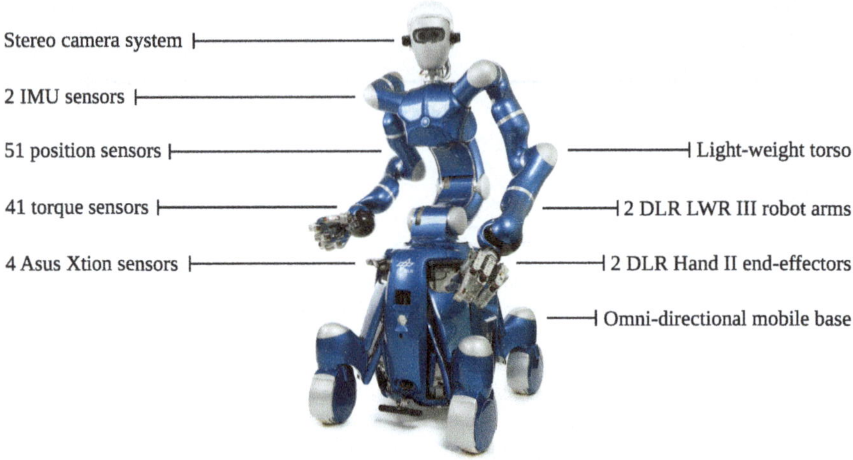

Fig. 2.2 Sensors and actuators of the humanoid robot Rollin' Justin

adult. It can reach objects at up to a height of up to 2.7 m, yet pass through narrow doorways as little as 0.9 m wide. This enables the robot to operate in human environments and manipulate the objects designed for humans, such as doors, furniture, tools, and objects of daily living. The robot can function as a completely standalone unit, and can be operated without any cables as it is equipped with a battery for about one hour of operation depending on the workload. All electronic components as well as the computers to control the robot are integrated into the housings. The interface to the robot is based on a Wi-Fi link, which allows to command the robot by means of mobile devices such as tablet computers and smartphones. In addition to this, the robot can be commanded via a speech recognition interface.

The mechanical construction of the robot integrates the concepts of modularity and integrated design. The arms of the robot are modified DLR light-weight robots (LWR) of the third generation (Hirzinger et al. 2002). The zero positions of the "elbow" joint is turned to more closely resemble that of a human arm. In addition, the "wrist" joint is modified to exhibit the properties of a hinge joint in order to reflect human wrist kinematics. The arms consist of seven *Degrees of Freedom (DOF)* each, arranged in a roll-pitch-roll-pitch-roll-pitch-pitch order. Each arm weighs about 14 kg and is able to lift 14 kg during slow motions, and about 7 kg at maximum velocity. The hands of Rollin' Justin are the four-fingered DLR II Hands (Butterfaß et al. 2001) with three actuated DOF and one coupled last joint per finger. An additional actuator allows to rearrange the palm configuration and change the position of the thumb and the ring finger to facilitate grasping of smaller objects. The arms of Rollin' Justin are mounted on a light-weight torso with three actuated DOF and a kinematically coupled fourth joint in the chest to force an upright position of the head. The head rests on a pan-tilt unit and is currently equipped with a stereo camera pair to perceive

Table 2.1 Overview of the actuators of Rollin' Justin

Actuator	Active DOF	Joint sensors
Torso	3	Position, torque
Arms	2×7	Position, torque
Hands	2×12	Position, torque
Neck	2	Position
Platform	8	Position
Total	51	

distant objects, an *Asus Xtion* motion sensing device for close range operation, and an *Inertial Measurement Unit (IMU)* for equilibrioception. Altogether the upper body weighs about 45 kg. All joints incorporate absolute position sensors. All upper body joints except the two neck joints are equipped with link-side torque sensors which enable advanced control strategies at 1 kHz for compliant whole-body manipulation to be detailed in Chap. 6. The actuators and their sensors are listed in Table 2.1.

The mobile base allows Rollin' Justin to reposition it self and extend the workspace over wide areas (Fuchs et al. 2009). It is mainly constructed from commercial off-the-shelf parts and contains the computers, the battery, and the majority of all electronic components. The base houses one more IMU and three additional Xtion sensors to perceive the immediate vicinity visually and avoid collision with obstacles. The footprint of the base can be extended by means of a parallel mechanisms. It can be locked to ensure stability during manipulation and a small baseline for navigation. The parallel mechanism is not actuated. Instead it can be reconfigured by means of active driving with the steerings and the wheels. A passive suspension system allows Rollin' Justin to overcome small obstacles. The nonholonomic platform can realize omni-directional maneuvers in the two translational directions and the rotation around the vertical axis. Altogether, Rollin' Justin is most suitable to accomplish the tasks that are discussed within this work.

2.3 Motion Planning and Generation

Executing motions with arms, hands, and the rest of the body is a natural procedure for human beings that does not need any conscious mental computation. Even sophisticated operations including the manipulation of multiple objects at once while avoiding obstacles in the environment are possible in no time. In contrast, it is a non-trivial task for a robot to plan and generate coordinated joint motions. In robotics, this problem is described by the term *motion planning*. Motion planning is typically concerned with the search for a collision-free path in the joint space, i. e. the *configuration space* (C_{space}) of a robot, where motions are planned to transit in between objects or transfer objects from one position to another. However, in comparison to this, wiping motions and other compliant manipulation tasks require paths that deliberately go into contact with the environment. While collision-free motion planning is a well studied problem, this work develops the methodologies to incorporate deliberate contact as well. A survey on the most common motion planning techniques is given by Latombe (1990). Traditional motion planning algorithms can be divided into *local* methods and *global* methods.

Local motion planning methods are typically based on gradient descent on artificially defined potential fields (Barraquand et al. 1992). While there are several derivatives of this approach, the basic concept remains the same. In short summary, the initial position q_{init} is defined as the source and the goal position q_{goal} is defined as a sink with attractive potential. Obstacles are modeled as repulsive potential fields such that they are avoided by the manipulator. Potential field methods require low computational effort. However, due to their local nature these methods are vulnerable to getting stuck in local minima.

Global motion planning algorithms attempt to find a feasible plan in the entire search space. That is, the entire C_{space} is explored to avoid obstacles and narrow passages. The most well known sub-class of global motion planning algorithms is defined by sampling based strategies. These strategies are often biased to explore the free space as fast as possible. Two popular examples are given by the *Probabilistic Roadmaps (PRM)* approach by Kavraki et al. (1996) and *Rapidly Exploring Random Trees (RRT)* introduced by LaValle (1998). PRMs define a path in two steps. First, a roadmap is generated by randomly sampling configurations in the C_{space} and connecting them locally. Second, the roadmap is searched for the shortest path between the start and the goal configuration by performing a graph search. This type of motion planning is especially useful for recurring tasks in static domains. A frequent adaption of the algorithm is a lazy evaluation of the roadmap, which allows to adapt quicker on changes in the environment (Bohlin and Kavraki 2000). Similar to PRMs, the RRT algorithm samples random configuration in the C_{space}. However, the sampled nodes form a search tree rather then a map. New configurations are sampled within a certain distance to the tree and directly connected to the closest leaf node. One of the most common variations of the RRT algorithm is the bi-directional search for a feasible path (Kuffner and LaValle 2000). The goal-oriented nature of RRTs make them perfect for onetime problems. As such they are better suited for service robotics applications where the environment is less defined than in industrial settings.

Another common way to generate motions is the computation of *Dynamic Movement Primitives (DMP)* (Schaal 2006). DMPs describe motions based on differential equations. The parameterization of these equations (i. e. the shape of the trajectory) is most efficiently conducted by means of supervised learning (Schaal et al. 2005). Instead of generating new motions for every problem from scratch, it is proposed by Stulp et al. (2012) to recall previously learned DMPs and optimize the shape *and* the goal parameters likewise, in order to adapt effectively to new problem instances.

The problems discussed in this work are highly versatile w. r. t. their domains, the involved objects, the number of active DOF, and the desired goal of the actions. Accordingly, it is most desirable to generate motions as flexible as possible. For this reason, the RRT algorithm is applied to generate free space motions in the remainder of this manuscript, if not otherwise stated. In particular, the implementation of the bi-directional RRT algorithm provided with the *Open Robotics Automation Virtual Environment (OpenRAVE)* is utilized as standard motion planner (Diankov 2010). Whenever contact is desired, the motion planning strategies developed within Chap. 5 are utilized.

2.4 Symbolic Planning and Logic Programming

Generic problem solving is a fundamental requirement to fully automate the operation of robots. It covers the problem of finding a suitable set of actions or operations to reach a desired goal state in a certain domain. As such, the field of automated planning was established as one of the first topics of AI. The field discusses the problem, of how an autonomous agent can generate a sequence of actions that lead to a certain goal-state, i. e. a desired effect to the world. This problem is typically of purely descriptive nature, where the described actions are treated as black-box that produce the described effects without any further geometric reasoning. An overview on this topic is provided by Ghallab et al. (2004). In the context of wiping tasks, this topic is considered with the integration of wiping actions into the overall reasoning framework. In particular, it is discussed how the effects of wiping motions can be represented in such a way that a symbolic planner can schedule the appropriate wiping action for a given problem instance.

Sometimes referenced to as *symbolic planning* (or simply planning), the field emerged from the *Stanford Research Institute Problem Solver (STRIPS)* (Fikes and Nilsson 1971). STRIPS describes both, a classical planning approach and a formal language to describe planning domains. A STRIPS domain is defined by an *initial state*, a *goal state*, and *actions* that are described by means of logic conjunctions and disjunctions of *preconditions* and *postconditions*, i. e. *effects*. The STRIPS language is the predecessor for many action languages to express problems in automated planning. The most common action language nowadays is the *Planning Domain Definition Language (PDDL)* (Ghallab et al. 1998), which is used in this work to describe the symbolic domains and problem definitions of the tasks investigated.

Classical planning has evolved into many directions, including graph planning and heuristic search, which are subject to active research. One graph planning approach of particular interest for robotics research is the concept of *Hierarchical Task Networks (HTN)* (Sacerdoti 1975). HTNs define the dependency between actions in form of a network. Similar to classical planning, an initial state is the starting point to solve a *set of tasks* in a predefined domain. In order to reach a desired goal state in this domain, it is proposed to recursively decompose a task into subtasks until a primitive task is reached that can be resolved by a planning operator. As an example, HTN planning finds application in human-robot interaction planning (Alami et al. 2006). Another well adopted planning scheme is given with heuristic progression planning (Bonet and Geffner 1999). These types of planners apply heuristic search strategies to find a solution for a desired goal state. Depending on the problem different heuristics may perform better then others. As an example, the *Fast Downward* planner proposed by Helmert (2006) resolves plans by first applying heuristics to top-level goals, to eventually proceed *downward* in the task hierarchy. The Fast Downward planner is applied as an out-of-the-shelve solution in the planning framework introduced with Chap. 5. As the planner is based on PDDL syntax, it can easily be exchanged in the future with different symbolic planners to extend the capabilities of the framework and compare different symbolic planners w. r. t. their performance.

A topic related to symbolic planning is found with *logic programming* (Lloyd 2012). Logic programming (sometimes associated with the research on computational logic) describes a programming paradigm that aims to express facts and rules in a logical form. It is possible to solve planning problems by defining the rules that declare how plans can be constructed from a sequence of actions. However, logic programming languages are much richer in their nature as they build on generic rules of deductive problem solving. Furthermore, logic programming languages facilitate exhaustive search as they possess intrinsic backtracking capabilities. As such, logic programming languages are predestined to query information from large datasets. One of the first logic programming languages is *Prolog* (Colmerauer and Roussel 1996). Developed in the early 1970s, Prolog has become the most popular language for logic planning, evident in its ever growing community. In this work, Prolog is not used to solve planning problems. However, Prolog is the language of choice for the openEASE reasoning framework (Beetz et al. 2015). The openEASE framework allows AI and robotics researchers to query big data from episodic memories of robot activities, where Prolog is used to design queries as it is done in Chap. 7. Accordingly, it is fundamental to this work. For a detailed introduction to Prolog programming, please refer to Sterling and Shapiro (1994).

2.5 Knowledge Representation and Reasoning

The field of knowledge representations and the accompanied reasoning methods deal with the question of how information can efficiently be represented for the use in computational models (Sowa et al. 2000). The models are constructed from logic relations to eventually create an *ontology* to better understand a certain domain. The creation of ontologies is the attempt to interpret single entities, their classification and (hierarchical) relations, and the singularization of similarities and differences. In the case of robotic manipulation, this involves the representation of task knowledge in form of objects, actions, motions, and effects. A robot with access to such information is potentially able to make intelligent decisions w. r. t. the current environmental situation. Accordingly, a classification of compliant manipulation tasks is conducted in Chap. 3, in order to gain a better understanding of the investigated problem of compliant manipulation. Based on this, representations to reason about wiping tasks are introduced in Chap. 4. The particular representation for wiping tasks includes the motion and the related effects of wiping actions.

Formal representations are thereby important to generalize the reasoning process toward versatile manipulation. A generic method to represent ontologies is developed with the *Web Ontology Language (OWL)* (McGuinness et al. 2004). OWL describes a formal method to link the terms of a certain domain by means of their semantic relation. As an example, OWL is used by the *KnowRob* system (Tenorth and Beetz 2009) to describe the domain of robotic manipulation tasks. The *Cognitive Robot Abstract Machine (CRAM)* (Mösenlechner and Beetz 2011) provides the means to reason about the knowledge within this domain and ground the semantic information of stored robot plans to executable commands. Besides this, representing knowledge is closely coupled to the fields of symbolic planning and logic programming, introduced in Sect. 2.4. As already stated, both PDDL and Prolog may be used to define the relation between actions, objects, and effects, i.e. an ontology. However, the stored information would be of symbolic nature only. There is no geometric information available that could be used to transfer the descriptive action definition into executable robot code. Nonetheless, PDDL is most common for planning methods and thus favorable w.r.t. the scope of this manuscript, which is concerned with the automated planning of everyday manipulation tasks. Accordingly, Chap. 4 details the development of a PDDL-based knowledge representation, which includes geometric process models that output executable robot operations. The methods to facilitate this reasoning shall be introduced with Chap. 5.

References

Gerd, Hirzinger, Norbert Sporer, Alin Albu-Schäffer, M. Hähnle, Rainer Krenn, A. Pascucci, and Manfred Schedl. 2002. DLR's Torque-Controlled Light Weight Robot III-Are we Reaching the Technological Limits Now? In *Proceedings of the IEEE International Conference on Robotics and Automation (ICRA)*, vol. 2, 1710–1716. IEEE.

Iwata, Hiroyasu, and Shigeki Sugano. 2009. Design of Human Symbiotic Robot TWENDY-ONE. In *Proceedings of the IEEE International Conference on Robotics and Automation (ICRA)*, 580–586. IEEE.

Sakagami, Yoshiaki, Ryujin Watanabe, Chiaki Aoyama, Shinichi Matsunaga, Nobuo Higaki, and Kikuo Fujimura. 2000. The Intelligent ASIMO: System Overview and Integration. In *IEEE/RSJ International Conference on Intelligent Robots and Systems (IROS)*, vol. 3, 2478–2483. IEEE.

Diftler, Myron A., J.S. Mehling, Muhammad E. Abdallah, Nicolaus A. Radford, Lyndon B. Bridgwater, Adam M. Sanders, Roger Scott Askew, D. Marty Linn, John D. Yamokoski, F.A. Permenter, et al. 2011. Robonaut 2-the First Humanoid Robot in Space. In *Proceedings of the IEEE International Conference on Robotics and Automation (ICRA)*, 2178–2183. IEEE.

Fitzgerald, Conor. 2013. Developing Baxter. In *Proceedings of the IEEE International Conference on Technologies for Practical Robot Applications (TePRA)*, 1–6. IEEE.

Englsberger, Johannes, Alexander Werner, Christian Ott, Bernd Henze, Maximo A. Roa, Gianluca Garofalo, Robert Burger, Alexander Beyer, Oliver Eiberger, Korbinian Schmid, et al. 2014. Overview of the Torque-controlled Humanoid Robot TORO. In *IEEE-RAS International Conference on Humanoid Robots*, 916–923.

Borst, Christoph, Thomas Wimböck, Florian Schmidt, Matthias Fuchs, Bernhard Brunner, Franziska Zacharias, Paolo Robuffo Giordano, Rainer Konietschke, Wolfgang Sepp, Stefan Fuchs, et al. 2009. Rollin' Justin—Mobile Platform with Variable Base. In *Proceedings of the IEEE International Conference on Robotics and Automation (ICRA)*, 1597–1598.

Albu-Schäffer, Alin, Christian Ott, and Gerd Hirzinger. 2007. A Unified Passivity-Based Control Framework for Position, Torque and Impedance Control of Flexible Joint Robots. *The International Journal of Robotics Research* 27 (1): 23–39.

Ott, Christian. 2008. *Cartesian Impedance Control of Redundant and Flexible-Joint Robots*, vol. 49. Berlin: Springer Publishing Company. Springer Tracts in Advanced Robotics.

Dietrich, Alexander, Thomas Wimböck, Alin Albu-Schäffer, and Gerd Hirzinger. 2012. Reactive Whole-Body Control: Dynamic Mobile Manipulation using a Large Number of Actuated Degrees of Freedom. *IEEE Robotics & Automation Magazine* 19 (2): 20–33.

Nilsson, Nils J. 1984. Shakey the Robot. Technical report, DTIC Document.

Stulp, Freek, Evangelos A. Theodorou, and Stefan Schaal. 2012. Reinforcement Learning with Sequences of Motion Primitives for Robust Manipulation. *IEEE Transactions on robotics* 28 (6): 1360–1370.

Billard, Aude, Sylvain Calinon, Ruediger Dillmann, and Stefan Schaal. 2008. Robot Programming by Demonstration. In *Springer Handbook of Robotics*, 1371–1394. Springer.

Tenorth, Moritz, and Michael Beetz. 2009. KnowRob—Knowledge Processing for Autonomous Personal Robots. In *Proceedings of the IEEE/RSJ International Conference on Intelligent Robots and Systems (IROS)*, 4261–4266. IEEE.

Beetz, Michael, Moritz Tenorth, and Jan Winkler. 2015. Open-EASE—A Knowledge Processing Service for Robots and Robotics/AI Researchers. In *IEEE International Conference on Robotics and Automation (ICRA)*, 1983–1990.

Ott, Christian, Oliver Eiberger, Werner Friedl, Berthold Bäuml, Ulrich Hillenbrand, Christoph Borst, Alin Albu-Schäffer, Bernhard Brunner, Heiko Hirschmüller, Simon Kielhöfer, et al. 2006. A Humanoid Two-Arm System for Dexterous Manipulation. In *Proceedings of the IEEE-RAS International Conference on Humanoid Robots (ICHR)*, 276–283. IEEE.

Borst, Christoph, Christian Ott, Thomas Wimbock, Bernhard Brunner, Franziska Zacharias, Berthold Bauml, Ulrich Hillenbrand, Sami Haddadin, Alin Albu-Schaffer, and Gerd Hirzinger. 2007. A Humanoid Upper Body System for Two-handed Manipulation. In *Proceedings of the IEEE International Conference on Robotics and Automation (ICRA)*, 2766–2767. IEEE.

Butterfaß, Jörg, Markus Grebenstein, Hong Liu, and Gerd Hirzinger. 2001. DLR-Hand II: Next Generation of a Dextrous Robot Hand. In *Proceedings of the IEEE International Conference on Robotics and Automation (ICRA)*, 109–114.

Fuchs, Matthias, Christoph Borst, P. Robuffo Giordano, Andreas Baumann, Erich Kraemer, Jörg Langwald, Robin Gruber, Nikolaus Seitz, Georg Plank, Klaus Kunze, et al. 2009. Rollin' Justin—Design Considerations and Realization of a Mobile Platform for a Humanoid Upper Body. In *Proceedings of the IEEE International Conference on Robotics and Automation (ICRA)*, 4131–4137. IEEE.

Latombe, Jean-Claude. 1990. *Robot Motion Planning*. Springer.

Barraquand, Jerome, Bruno Langlois, and J.-C. Latombe. 1992. Numerical Potential Field Techniques for Robot Path Planning. *IEEE Transactions on Systems, Man, and Cybernetics*, 22 (2): 224–241.

Kavraki, Lydia E., Petr Svestka, Jean-Claude Latombe, and Mark H. Overmars. 1996. Probabilistic Roadmaps for Path Planning in High-dimensional Configuration Spaces. *IEEE Transactions on Robotics and Automation* 12 (4): 566–580. ISSN 1042-296X.

LaValle, Steven M. 1998. Rapidly-exploring Random Trees: A New Tool for Path Planning.

Bohlin, Robert, and Lydia E. Kavraki. 2000. Path Planning using Lazy PRM. In *Proceedings of the IEEE International Conference on Robotics and Automation (ICRA)*, vol. 1, 521–528. IEEE.

Kuffner, James J., and Steven M. LaValle. 2000. RRT-Connect: An Efficient Approach to Single-Query Path Planning. In *Proceedings of the IEEE International Conference on Robotics and Automation (ICRA)*, vol. 2, 995–1001. IEEE.

Schaal, Stefan. 2006. Dynamic Movement Primitives—A Framework for Motor Control in Humans and Humanoid Robotics. In *Adaptive Motion of Animals and Machines*, 261–280. Springer.

Schaal, Stefan, Jan Peters, Jun Nakanishi, and Auke Ijspeert. 2005. Learning Movement Primitives. In *The 11th International Symposium on Robotics Research.*, 561–572. Springer.

Diankov, Rosen. 2010. *Automated Construction of Robotic Manipulation Programs*. PhD thesis, Carnegie Mellon University, Robotics Institute.

Ghallab, Malik, Dana Nau, Paolo Traverso. 2004. Automated Planning: Theory and Practice. Morgan Kaufmann.

Fikes, Richard E., and Nils J. Nilsson. 1971. STRIPS: A New Approach to the Application of Theorem Proving to Problem Solving. *Artificial Intelligence* 2 (3–4): 189–208.

Ghallab, Malik, Adele Howe, Dave Christianson, Drew McDermott, Ashwin Ram, Manuela Veloso, Daniel Weld, and David Wilkins. 1998. PDDL—The Planning Domain Definition Language. *AIPS98 Planning Committee* 78 (4): 1–27.

Sacerdoti, Earl D. 1975. A Structure for Plans and Behavior. Technical report, DTIC Document.

Alami, Rachid, Aurélie Clodic, Vincent Montreuil, Emrah Akin Sisbot, and Raja Chatila. 2006. Toward Human-Aware Robot Task Planning. In *AAAI Spring Symposium: To Boldly go where No Human-robot Team has Gone Before*, 39–46.

Bonet, Blai, and Hector Geffner. 1999. Planning as Heuristic Search: New Results. In *European Conference on Planning*, 360–372. Springer.

Helmert, Malte. 2006. The Fast Downward Planning System. *Journal of Artifcial Intelligence Research* 26: 191–246.

Lloyd, John W. 2012. *Foundations of Logic Programming*. Springer Science & Business Media.

Colmerauer, Alain, and Philippe Roussel. 1996. The Birth of Prolog. In *History of Programming Languages II*, 331–367. ACM.

Sterling, Leon, and Ehud Y. Shapiro. 1994. *The Art of Prolog: Advanced Programming Techniques*. MIT Press.

Sowa, John F., et al. 2000. *Knowledge Representation: Logical, Philosophical, and Computational Foundations*, vol. 13. MIT Press.

McGuinness, Deborah L., Frank Van Harmelen, et al. 2004. OWL Web Ontology Language Overview. *W3C Recommendation*, 10 (10): 2004.

Mösenlechner, Lorenz, and Michael Beetz. Parameterizing Actions to Have the Appropriate Effects. In *IEEE/RSJ International Conference on Intelligent Robots and Systems (IROS)*, 4141–4147. IEEE.

Chapter 3
Classifying Compliant Manipulation

This chapter investigates the properties of compliant manipulation tasks in general, and wiping tasks in particular, in order to gain a better understanding of the problem. As a result, a taxonomy shall be presented to build the foundation for the development of suitable representations to plan, execute, and interpret wiping tasks. In particular, a two-step classification is proposed to first cover the symbolic nature of the contact for compliant manipulation task in general, and subsequently classify wiping tasks according to the geometric goal state of the medium. The developed taxonomies shall highlight the differences and emphasize similarities between different compliant manipulation actions. This allows software engineers to develop generalized robot actions beyond obvious similarities and reduce the development effort significantly, especially when the actions are developed independently of any robot and applicable to arbitrary domains.

Chapter 3 reviews the available classification approaches in order to develop a more suitable classification matching the needs of compliant manipulation in robotics. Section 3.1 reviews the literature of related classifications for robot manipulation and makes an attempt to categorize wiping tasks accordingly. Section 3.2 develops a novel taxonomy tailored according to the needs of compliant manipulation tasks and their semantics. It is discussed which implications the categorization of a task has for the reasoning procedure as well as the applied control concepts. Finally, wiping actions are investigated in more detail in Sect. 3.3 to develop the nine fundamental wiping actions.

The findings in this chapter are published in (Leidner et al. 2016b) and (Leidner et al. 2015a). In particular, a preliminary classification of wiping tasks was conducted in (Leidner et al. 2016b). Based on this, an elaborate investigation on applied compliance in daily environments has led to the final taxonomy of compliant manipulation tasks (Leidner et al. 2015a) as utilized to motivate the main concepts developed in this manuscript.

D. S. Leidner, *Cognitive Reasoning for Compliant Robot Manipulation*, Springer Tracts in Advanced Robotics 127,
https://doi.org/10.1007/978-3-030-04858-7_3

3.1 Manipulation Taxonomies in Robotics Research

In the past, several manipulation taxonomies have been designed to categorize actions w. r. t. different properties. In order to design a suitable taxonomy, one may apply an object-centric approach (Wörgötter et al. 2013), motion-based viewpoints (Bloomfield et al. 2003), or an effect-based point of view (Vukobratović and Potkonjak 1999). In addition, the available literature discusses classifications based on finger positions (Cutkosky 1989), relative hand motions (Bullock et al. 2013), or geometric dimensions (Morrow and Khosla 1997). These taxonomies mostly rely on anthropomorphic kinematics and neglect the effect to the environment. Accordingly, they are hardly applicable as blueprint to design robot independent actions. This observations goes hand in hand with the traditional development of robot actions. These actions specify the desired effect implicitly based on the experience of expert users. This renders the robot clueless about the purpose of its motions, such that it is unaware of the resulting changes to the world. In contrast, AI reasoning techniques show how the effect of an action can be properly taken into account by explicitly specifying preconditions and postconditions, i. e. effects. A classification that incorporates both viewpoints would enable the development of generic actions that consider geometric and semantic features likewise. Even though the available classifications are not always designed w. r. t. robotic applications they have been widely adapted by robotics researchers. Therefore, this section reviews the literature on manipulation taxonomies and try to make an attempt to categorize compliant manipulation tasks respectively.

3.1.1 Related Work

The available attempts to classify manipulation can be separated into *Grasp Taxonomies* and *Manipulation Taxonomies*. As a third class, this manuscript lists *Hybrid Taxonomies* that are derived from the two main classes.

Grasp Taxonomies such as the taxonomy by Kapandji and Honoré (1970), were originally developed to evaluate the grasping capabilities of human hands. In particular, Kapandji and Honoré utilized their taxonomy to evaluate the success of hand-surgeries. One of the most well-known taxonomies in robotics is probably the taxonomy of Cutkosky (1989). It consists of 16 grasp types ordered in a hierarchical tree based on an elaborate study on a set of machining tasks. The main classes derived from this study constitute *power grasps* and *precision grasps*. Based on this and many other taxonomies, Feix et al. (2009) developed a comprehensive grasp taxonomy that captures over 33 grasp classes. Grasp taxonomies have been widely adopted in robotics research. They have been used to plan grasping procedures as shown by Stansfield (1991). Grebenstein et al. (2011) describe how grasp taxonomies can be used to develop anthropomorphic robot hands. Chalon et al. (2014) determine the performance of anthropomorphic robotic hands. The problem with pure grasp taxonomies is that they are not applicable to develop robot actions. That is, a grasp cannot directly and exclusively be related to a certain type of action. Moreover, the available grasp taxonomies are based on the kinematics of anthropomorphic hands and therefore not suited for generic robotic manipulators. As an example, industrial manipulators often only provide two-jaw grippers, or other exchangeable tools.

Manipulation Taxonomies differ from grasp taxonomies as they do not solely rely on the configuration of a hand, but also consider the environment, i. e. the objects that are manipulated. This is an important difference, since a certain object may be handled with many different hand poses and one hand pose can be applied to several objects. Only for some cases it is possible to infer the intended action based on how an object is grasped (e. g. holding a screwdriver on the handle to turn a screw). While Grasp Taxonomies are of purely geometric nature (i. e. they do only refer to actions or effects to describe example cases), manipulation taxonomies may be sub-divided into geometric and semantic classifications:

Geometric Manipulation Taxonomies consider spatial relations between objects and the environment over time. An object-centric classification has been conducted by Morrow and Khosla (1997). The authors specify a set of primitive actions in Cartesian space based on geometric constraints. They propose to formulate more complex actions by concatenating several primitive actions. This approach allows to schedule robot actions in an abstract form. However, the level of abstraction is not suited for automated planning where semantic information is necessary. Bloomfield et al. (2003) mainly classify haptic actions w. r. t. forces and torques applied to the operators hand. The taxonomy is capable of describing common situations observed in everyday household environments and industrial machining tasks. However, as these parameters depend on the relative position of the object to the hand, this taxonomy is hardly applicable to arbitrary robotic manipulators. The hand-centric taxonomy introduced by Bullock et al. (2013) classifies the relative motion of human hands w. r. t. the grasped or touched object during a certain task. Similarly to Cutkosky, Bullock et al. define a hierarchical tree structure to distinguish different classes of manipulation. The two main classes are *prehensile* and *non-prehensile* manipulation. Based on their findings, Bullock et al. (2013) investigate in-hand manipulation in more detail. Eventually, they develop a taxonomy of translations and rotations of the object in hand.

Semantic Manipulation Taxonomies mostly neglect the numeric meaning of geometric relations. Instead, they consider motions and actions from an descriptive point of view. Wörgötter et al. (2013) focus on assembly and dis-assembly tasks. They classify actions according to the relation of arbitrary objects in space and time. They derive a semantic classification for the investigated domain. The main set of tasks consists of *hand-only actions*, *separation actions*, and *release determined actions*. The authors outline how their taxonomy can be used to schedule relatively complex assembly tasks. However, it is limited to this domain. Vukobratović and Potkonjak (1999) propose a taxonomy based on the semantic description of contact situation between manipulated objects and the environment. They argue that such a taxonomy allows to make assumptions about the dynamics of the tasks.

Among others, the taxonomy comprises *reaction force*, *impact*, *contact friction*, and *deformation*. Furthermore, the authors suggest to develop a unified model to cover all effects mentioned. While, it is arguable if a single procedure can result in actions with performant task execution for every case the idea of generic process models is similar to the approach applied in this work.

Hybrid Taxonomies are not too common in the literature. Only recently, Liu et al. (2016) proposed a hybrid taxonomy. The authors attempt to combine traditional grasp taxonomies with manipulation taxonomies by observing grasps in action. The authors classify finger joint positions in relation to the action that is performed by annotating the grasps with common English words. Accordingly, the taxonomy applies a mixed geometric and semantic point of view. An interesting aspect of this work is the insight that the grasp parameters for grasps in action are similar to the parameters that are required to define impedance behavior (i. e. the motion, force, and stiffness properties) of robotic manipulators. In fact, this observation matches the motivation for this manuscript outlined in Sect. 1.3.

3.1.2 Investigating Applicable Taxonomies

The classifications related to research on robotic manipulation are listed in Table 3.1. The table is sorted according to the classification type and their relation (i. e. symbolic or geometric). Not all taxonomies are applicable to classify compliant manipulation tasks. Especially the listed grasp taxonomies are unable to make statements about the purpose of actions. The taxonomy by Wörgötter et al. (2013) is limited to the domain of assembly and disassembly tasks and therefore not applicable as well. The remaining taxonomies are investigated closer w. r. t. the categorization of actions that requires compliant behavior. The benefits and limitations of the corresponding classifications are extracted, such that the requirements for a new taxonomy (matching the needs of compliant manipulation with application to automated planning) can be formulated.

Compliant manipulation in Morrow and Khosla (1997)

The classification by Morrow and Khosla (1997) is an early attempt to "*build a richer library of robot capabilities in the manipulation domain*" (Morrow and Khosla 1997). The authors explicitly mention the deep integration of sensing capabilities into robot programming primitives. With this in mind, the proposed taxonomy describes robot actions as geometric primitives that are constrained in motion. For example, a cylindric peg-in-hole task can be broken down to one possible transitional motion in parallel to the peg/hole and one unconstrained rotation around the main axis of the peg/hole. Similarly to the motion constraints, the primitives define which force interactions result in undesired effects (e. g. jamming), and in which dimensions the robot shall comply in order to align with the task-space. By combining multiple primitives one can describe more complex skills. Developing actions based on this approach, a robot could be programmed to execute compliant manipulation tasks.

Table 3.1 Available manipulation taxonomies related to robotics research

Type	Publication	Relation	Applicable
Grasp	Kapandji and Honoré (1970)	Geometric	No
Grasp	Cutkosky (1989)	Geometric	No
Grasp	Feix et al. (2009)	Geometric	No
Manipulation	Morrow and Khosla (1997)	Geometric	Yes
Manipulation	Bloomfield et al. (2003)	Geometric	Yes
Manipulation	Bullock et al. (2013)	Geometric	Yes
Manipulation	Vukobratović and Potkonjak (1999)	Semantic	Yes
Manipulation	Wörgötter et al. (2013)	Semantic	No
Hybrid	Liu et al. (2016)	Mixed	Yes

However, the task description does not specify what the goal of this motion should be, since semantic information is missing for this taxonomy.

Compliant manipulation in Bloomfield et al. (2003)

The manipulation taxonomy developed by Bloomfield et al. (2003) was designed to describe haptic actions in disassembly tasks. The taxonomy mainly describes actions based on the forces and torques required to solve the task, where special interest is given to the direction in which force or torque is applied. The discrimination between forces and torques is an interesting aspect of compliant manipulation. Depending on the type of feedback, a skilled worker may alter the motion of a tool in order to accomplish a task. Yet, a generic application to compliant manipulation tasks is hardly possible as this feature is mainly dependent on how a tool is grasped. For example, grasping a brush at the handle results mainly in torques, whereas grasping it close to the bristles results in forces. However, a different grasp frame does usually not result in different effect.

Compliant manipulation in Bullock et al. (2013)

Bullock et al. propose a hand-centric classification of manipulation tasks based on geometric relations. The authors attempt to explain the context of the manipulation w. r. t. the hand position relative to the environment, the motion relative to the environment, the configuration of the hand, and the applied force vector. In contrast to traditional grasp taxonomies, it is possible to query information beyond the hand configuration such as the relation to the manipulated object in terms of abstract symbols, such as whether there is motion at contact or not. The main drawback of this taxonomy is that there is no information provided on the purpose of the actions. Accordingly, it is not possible to plan goal-oriented actions based on the manipulation descriptions provided.

Compliant manipulation in Vukobratović and Potkonjak (1999)

Vukobratović and Potkonjak investigate the interaction between two dynamic systems in a general manner. Based on this, they develop an approach to model a robot interacting with the environment. The authors pay special attention on the effect of the contact. Among others, the discussed effects include forces and torques for rigid-body-contact, elastodynamics in contact zones, friction observed during contact situations, elastic deformation, and impact. A generalized control strategy for robot manipulation is developed based on the inherited models. In fact, the identified effects play a major role in compliant manipulation and influence the control strategy for robotic manipulators, respectively. Even though it is not directly possible to identify single actions based on these effects, it is possible to group classes of actions accordingly.

Compliant manipulation in Liu et al. (2016)

The taxonomy developed by Liu et al. (2016) is the result of careful observations of humans performing daily activities. During their research, the authors noticed that many grasps cannot exclusively be mapped to a certain task. To create a more meaningful classification, Liu et al. propose to augment this seemingly different grasps by means of narratives, such as the intended action together with the executed motion, the applied force, and specified stiffness. As a result, a combined taxonomy is developed comprising semantic information and geometric features. However, as with other grasp taxonomies, the proposed hybrid taxonomy is designed to match human-like kinematics and therefore hardly applicable to arbitrary robotic systems.

In conclusion, none of the proposed taxonomies is able to satisfy the requirements to describe compliant robotic manipulation with application to AI-based reasoning and automated planning. In particular, most taxonomies focus on geometric properties and ignore the abstract nature of the effect, i. e. the semantic meaning of the action. Vice versa, the taxonomies that describe semantic aspects neglect the geometric realization completely. The only taxonomy incorporating both aspects is proposed by Liu et al.. However, it is limited to human-like hand kinematics. A generic taxonomy has to specify geometric constraints in a more generic and abstract way, such that the execution of actions can be described independent of the user or robot performing the task. For this reason, this chapter develops a novel taxonomy on compliant manipulation incorporating robot actions w. r. t. semantic information and geometric features likewise.

3.2 The Compliant Manipulation Taxonomy

Recent advancements in mechanical robot design and control enable the systems to get in soft contact with their environment. However, the deployed robots can vary significantly. Therefore, it is necessary to develop reliable, flexible, and generic robot actions, independent of a robot specific kinematics (e. g. human hand kinematics).

One way to overcome this issue is to develop robotic manipulation actions from an object point of view rather than relying on the robot capabilities. To minimize the development effort and to create sustainable software modules, actions have to be arranged in a higher level of abstraction. Therefore, this work proposes to classify compliant manipulation tasks w. r. t. the effects to the world on a symbolic level of abstraction and not solely based on geometric features and low-level control properties. The classification terms are therefore developed based on the objects that participate in the task execution as it is visualized in Fig. 3.1. Instead of developing the taxonomy based on kinematics features, a symbolic analysis is conducted. The classification is not limited to a specific manipulator or hand. Similarly it is not distinguished between one or more manipulators.

The developed manipulation taxonomy is illustrated in Fig. 3.2. Similarly to the classifications of Cutkosky (1989) and Bullock et al. (2013), a hierarchical tree structure is applied to guide the reasoning process. Depending on the depth in the tree, a robot has to reason in more detail about the required parameters (see Sect. 3.2.3). The taxonomy characterizes actions in a descriptive manner to make it complementary to the action definition in automated planning (Ghallab et al. 2004): Each branch defines a symbolic classifier related to the *contact situation* between an object or tool and a target in the environment, without limiting how the contact is exerted in particular. In the follow-up, generic process models can be derived by sub-categorizing each leaf node as done in Sect. 3.3. Consequently, a concrete implementation based on the classification has to integrate the respective parameters accordingly.

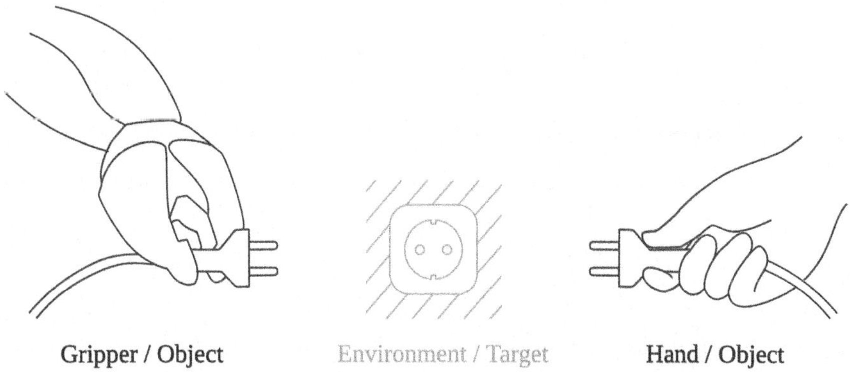

Fig. 3.1 An object-centric view is used to derive classification terms. Each classification term describes an abstract view on the contact situation between the two reference systems, i. e. the hand-object system colored in black and the environment colored in gray. It is thereby not of interest if the tool is guided by a robotic manipulator (left) or a human hand (right)

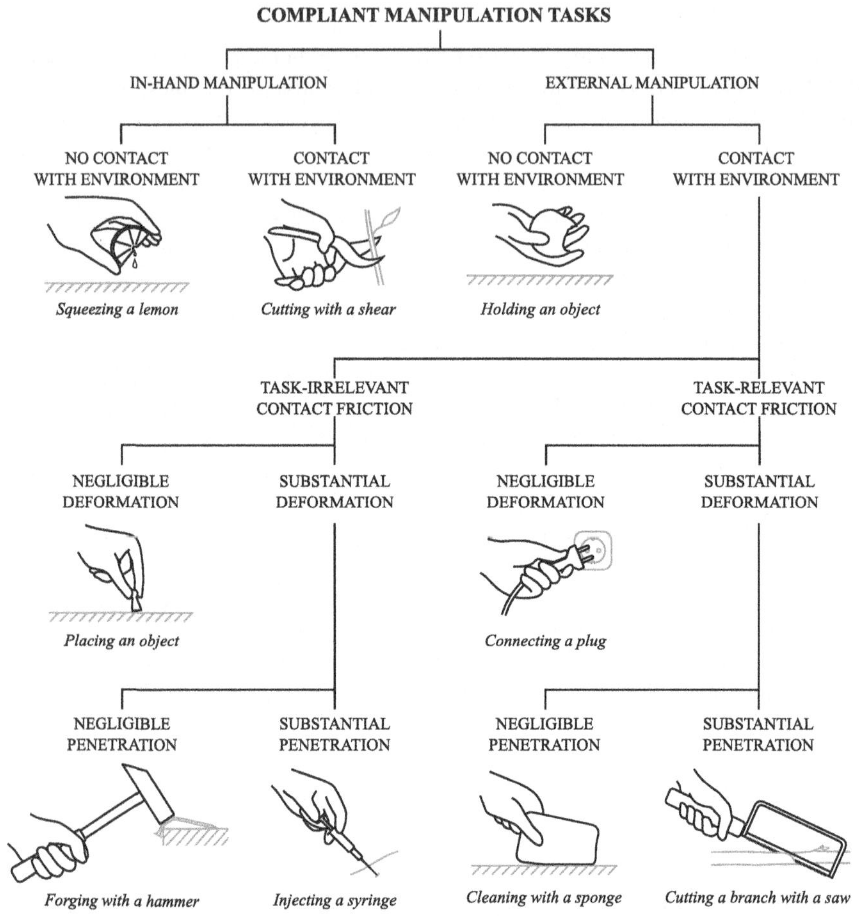

Fig. 3.2 Hierarchy of compliant manipulation tasks. For each leaf of the tree an example action is given. The complexity increases from top to bottom in number of parameters to be considered during the reasoning process. Note that neither hand kinematics, nor the hand posture, nor the contact between the hand and the grasped object are considered for classification

3.2.1 *Classification Terms*

This section describes the terms of the taxonomy. The classifiers are selected to categorize abstract action classes for automated planning. With each classifier, additional information is available to reason about the underlying effects to the environment, an adequate set of parameters, and the required control strategies (see Sect. 3.2.3). Since the approach is object-centric, each classifier is chosen w. r. t. the nature of the contact between the two reference systems, namely the *hand-object* system (black) and the *environment*, respectively the target object (gray).

In-Hand/External

The first classifier defines whether the force is exerted within the manipulator or hand (*in-hand*) or whether it originates from an external source in the environment (*external*). This term is related to the concept of *virtual linkage* introduced by Williams and Khatib (1993), which defines *internal forces* as forces within the grasp map of a manipulator.

Contact

This term defines whether the hand-object system makes deliberate contact with the environment (resulting in reaction force) and whether a proper set of compliant control parameters is crucial for the task performance. Force also originates without contact by acceleration (e. g. by gravity) of a mass, which also includes simply holding an object, for example.

Contact Friction

Contact friction is observed when an object is moved along the surface of another object in contact. Force has to be exerted in parallel the surface in order to overcome the friction. Friction may be irrelevant for certain manipulation tasks with small surfaces and is negligible then. It can also be a substantial part of an action with relevance for the task, as for example in cleaning tasks. Tasks involving friction are very sensitive to the direction of motion which may require different stiffness settings for individual Cartesian directions. As opposed to this, force is mainly exerted in other ways e. g. pressure, torque or upon impact, if friction is negligible.

Deformation

Non-rigid object manipulation requires to reason about the deformation of objects during the contact phase, but also afterwards if the deformation is persistent. Deformations occur either in the object to be manipulated, or the environment, respective, the target object. Geometric deformation may result in a different symbolic state for an object, e. g. it can be an indicator for damage. Note that all contacts results in deformations in microscopic scale which are neglected for the classification.

Penetration

Penetration occurs if the target structure can no longer accommodate the applied force. Penetration may lead to significant, irreversible alteration of the geometric state of an object including the full separation into multiple parts. This has to be considered on the symbolic level. Research on tool-soil interaction has proven that active variable force and stiffness is necessary while deliberately penetrating an object (soil) due to the varying resistance force (Leidner et al. 2015b). Note that any deformation can involve penetration in microscopic scale which is neglected for the classification.

Table 3.2 Manipulation task examples based on the developed classification terms

Category	Example Tasks
In Hand - No Contact	Squeezing a lemon, moving a pen in-hand
In Hand - Contact	Cutting with a shear, using a screwdriver
External - No Contact	Holding an object, lifting an object, throwing a ball
External - Contact - Irr. Friction - Neg. Deformation	Placing an object, touching objects, tactile exploration
External - Contact - Irr. Friction - Deformation - Neg. Penetration	Forging iron with a hammer, wind a rope or cable, folding paper
External - Contact - Irr. Friction - Deformation - Penetration	Injecting a syringe, stirring with a spoon, fork up food
External - Contact - Friction - Neg. Deformation	Connecting a plug, inserting a key, closing the lid of a pen
External - Contact - Friction - Deformation - Neg. Penetration	Cleaning a chalkboard, scrubbing a mug, ironing a shirt, sweeping the floor with a broom, painting a wall
External - Contact - Friction - Deformation - Penetration	Sawing a branch, cutting bread with a knife, excavate soil with a shovel

3.2.2 Example Actions

The provided example actions are representative for the combinations of the available parameters and discussed in the following. As the taxonomy is object-centric, the classification terms describe relations between the hand-object system and the environment. An excerpt of example actions are listed in Table 3.2. Note that the example actions describe snapshots of particular manipulation sequences for which the classification terms apply.

In Hand - No Contact

There is no contact between the hand-object system and the environment. Forces only occur intrinsic to the hand, as the object is manipulated in-hand.

In Hand - Contact

The hand-object system is in contact with the environment. The interaction with the environment propagates energy to the hand-object system. As a result, forces intrinsic to the manipulator occur.

External - No Contact

In contrast to in-hand manipulation, the main actor of external manipulation is the manipulator, i. e. the arm of the user/robot. Forces originate from accelerating a mass/inertia where the acceleration also involves gravity.

External - Contact - Irr. Friction - Neg. Deformation

Classical rigid contact with rigid body motions where force is transfered as e. g. pressure, torque, or impact. This action of task is typically executed without any notion of compliance.

External - Contact - Irr. Friction - Deformation - Neg. Penetration

Deformation occurs upon contact where friction is of lesser relevance or not salient to the task, i.e. pushing, pulling, bending, or hitting something.

External - Contact - Irr. Friction - Deformation - Penetration

A rigid object deliberately penetrates another object. Penetration is often considered an irreversible effect. The observed friction is significantly smaller than the resistance force and therefore irrelevant to the task execution.

External - Contact - Friction - Neg. Deformation

A rigid object is in physical contact with another rigid object, guided by the contours of the objects in a sliding motion. The sliding motion results in friction and jamming results in transverse forces.

External - Contact - Friction - Deformation - Neg. Penetration

A soft object is guided along the surface of a rigid object or vice versa. Also both objects can be soft. Usually the task involves a medium, which can be considered to be non-rigid in a macroscopic view. Most of these actions can be summarized as wiping tasks, to be detailed in Sect. 3.3.

External - Contact - Friction - Deformation - Penetration

Upon sliding contact, a rigid object penetrates the target deliberately. The penetration effect is irreversible. Eventually, penetration can result in the full separation of one object into two or more fragments.

3.2.3 Discussion

Each classification term provides additional information for the reasoning process and has to be considered during the task execution. According to the literature, especially the required control strategies have to be selected w. r. t. the nature of the contact. For example, squeezing a lemon mainly relies on compliant in-hand control strategies (Williams and Khatib 1993; Wimböck et al. 2011), whereas inserting a plug can be efficiently solved with Cartesian compliance at the end-effector, plus an appropriate strategy to prevent the plug from jamming (Stemmer et al. 2007). A task without penetration can exploit impedance-control with a dedicated stiffness to get in soft contact and wipe along a surface (Hogan 1987). If penetration is a substantial part of the action, force has to be adapted according to the penetration depth (Leidner et al. 2015b) or as soon as the penetration occurs (Xie et al. 2010). The varying requirements have to be taken into account by a properly parameterizable control framework (Dietrich et al. 2012). The control parameters depend thereby on the physical parameters (e. g. mass, inertia, or center of mass) of the objects involved in the task execution.

A generic classification can sometimes result in ambiguous cases where an action cannot be assigned to one particular class. For example, different tasks may or may not involve substantial friction depending on the actual state of the environment, e. g. pushing a door. In some cases inaccurate parameterization can lead to undesired effects. For example, if a sponge is used to clean a knife, too much force might lead to an entirely different action, namely cutting the sponge with the knife. It is notable that the objects involved in the task execution and the current environmental conditions always influence the task parameters, independent of any classification and should therefore always be considered during the reasoning process.

Furthermore, many tasks can only be described by combining multiple branches of the tree. Especially handling electric tools requires to activate a button in-hand while exerting force with the tool to the environment, such as drilling a hole with an electric drill. It is also possible that a sequence of actions describes a particular skill more sufficiently. For example, a peg-in-hole task may be defined as a combination of pushing, sliding and inserting, to make the task execution more robust as illustrated in Fig. 3.3. For this reason, a classification of robot skills has to comply with abstract effect descriptions to enable the automated planning of action sequences.

In general, the proposed classification can describe compliant manipulation tasks on a high level of abstraction. The described classification terms convey the appropriate control behavior for a classified task and are therefore valuable for the parameterization of robot skills. However, the classification terms are too abstract to directly implement concrete actions from it, since they only define the nature of the contact symbolically, and not how the contact is geometrically established. Therefore, a *two-step approach* is proposed to classify compliant manipulation tasks on both, symbolic and geometric levels of abstractions likewise. Accordingly, each leaf

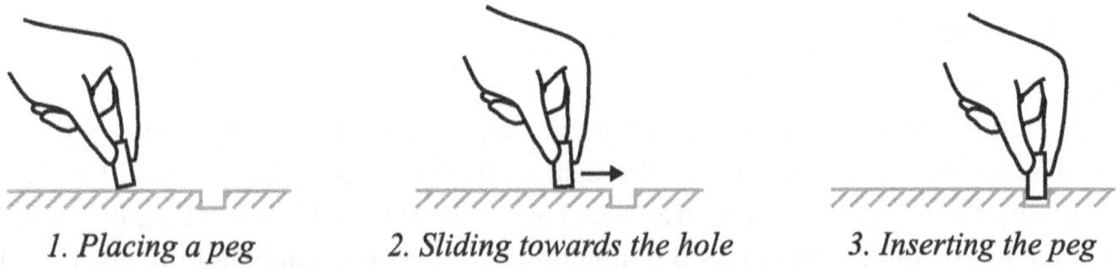

Fig. 3.3 Example of peg-in-hole insertion as combination of three actions

illustrated in Fig. 3.2 has to be revisited to extract similarities to derive generic process models. This sub-categorization is conducted for the category of *wiping tasks* in the following section.

3.3 Classification of Wiping Tasks

The main classification described in the previous section does not consider geometric parameters, but rather more abstract symbolic parameters. However, to create a connection between automated symbolic planning and geometric planning and parameterization of robotic manipulation tasks, the abstract definitions have to be grounded in robot operations. According to the analysis of Cakmak and Takayama (2013), this sub-classification is conducted for the category of the most frequent household chores. The authors conducted a survey to analyzed daily chore lists and found out that cleaning tasks are the most frequent household chores. In particular, 49.8% of the investigated tasks are related to *wiping surfaces* of *objects*, furniture, or rooms. Similar to the concept of this work, Cakmak and Takayama argue that "*tasks within a certain category exhibit similar structures that can be exploited while implementing robotic capabilities*" (Cakmak and Takayama 2013), where they explicitly address wiping and cleaning tasks.

In addition, a wide variety of cleaning-unrelated tasks based on the principle of wiping a surface can be observed, e. g. painting a wall. Besides the tool-surface contact, wiping tasks share another common component: the *medium* (particles/liquid *between* tool/surface, e. g. dust or paint). In summary, these tasks can be formulated as *guiding a tool along a target surface while maintaining contact to manipulate some sort of medium*. Wiping fits in the group of *External - Contact - Friction - Deformation - Neg. Penetration* w. r. t. the classification terms introduced in Sect. 3.2.2 (cp. Fig. 3.2). Accordingly, mainly friction and deformation have to be taken into account by the control strategy in contact. However, the actions differ in their *geometric process models*. For example, sweeping with a broom requires directed motions, while the trajectories in vacuuming may be random in general. The actions have to be distinguished carefully in order to deal with these differences in a generalized way. A sub-categorization is performed to extract geometric process models for the different types of wiping. De facto, a *two-step classification* is conducted to incorporate symbolic and geometric properties similarly.

3.3.1 Classification Terms

Similar to the contact analysis in the main classification, wiping tasks are not directly categorized w. r. t. numerical or geometric features, but rather classify them based on abstract terms to *implicitly* group actions with similar geometric structures. The

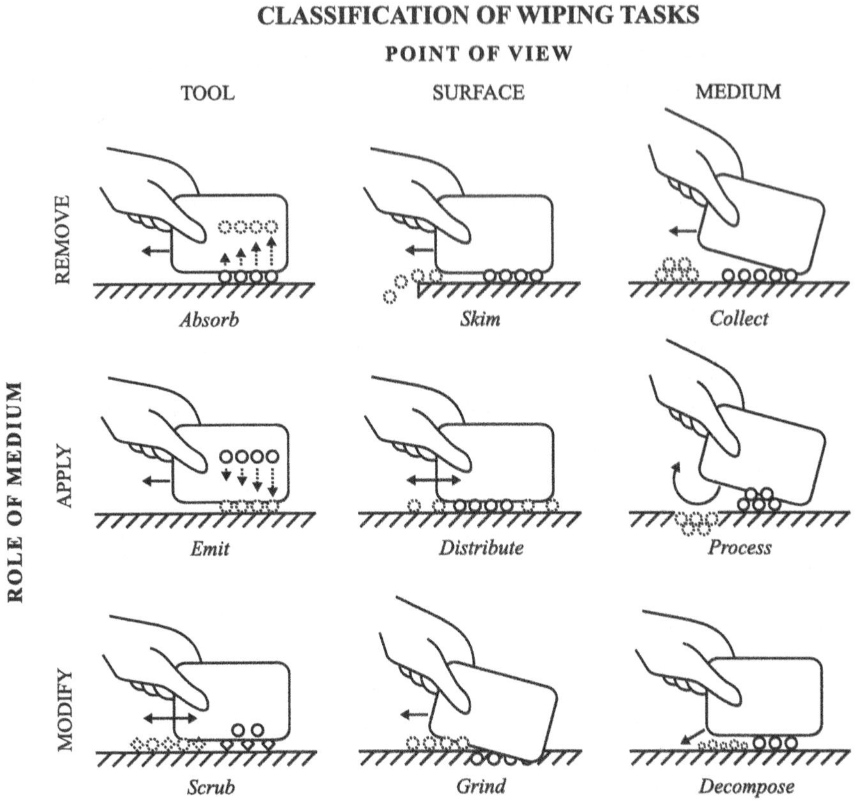

Fig. 3.4 Matrix classification of wiping tasks according to the tool, the surface, and the medium, versus the role of the medium. The tool (grasped by a human hand) is abstracted as a rectangle capable of all illustrated actions. The motion of the tool is indicated as a solid arrow. The surface is always shown on the bottom of the corresponding matrix cell. It might be flat as illustrated, curved, or of any other shape. The medium is shown shown in an abstract form in the initial state (solid circles) and in the goal state after the action is performed (dashed circles) where dashed arrows indicate the transition of the medium

tool-surface-medium tuple[1] is investigated, where the final geometric state of the medium corresponds to the desired symbolic goal state. Based on this role of the medium, individual process models can be derived. Nine action types related to specific sub-classes occurring in wiping tasks were identified. The actions are grouped in the procedures of *applying*, *removing*, and *modifying* the medium. Additionally, a *tool-centric view*, a *surface-centric view*, and a *medium-centric view* are applied to categorize the actions. The resulting matrix structure is illustrated in Fig. 3.4 and explained from the top left to the bottom right:

[1]The term tuple is generaly used to describe n-tuples in this manuscript.

Absorb

A medium is absorbed upon close vicinity to the tool. This may be caused by electrostatic force as known from dusting, an air draft from a vacuum, or capillary action as seen between a sponge and water. The effect is mostly unrelated to the direction of motion.

Skim

The final location of the medium is not of interest and is therefore illustrated as *skimmed from the surface*. Removing snow from a car window is considered as skimming. The direction of motion is defined by the individual geometric topology.

Collect

Collecting can be related to skimming. However, the medium has to be collected afterwards, e. g. to remove it accumulated. Representative actions are collecting leaves with a rake or sweeping up shards. The tool alignment w. r. t. the surface is crucial.

Emit

Emitting is the counterpart to absorbing. The medium is initially located on or in the tool and is applied to the surface as it is done for painting a wall, for example. Typically, the whole surface is involved.

Distribute

Distributing a medium is related to emitting a medium. However, the medium is already located on the surface. Applying shoe polish is such a task. The task trajectory is important to distribute the medium on the surface.

Process

Processing is a medium-centric action. The medium is used to alter the surface on purpose, as done with cement, for example. It is also possible that the surface is only used to directly manipulate the medium as done with cookie dough.

Scrub

Scrubbing merges an auxiliary medium with an unwanted medium (e. g. detergent and dirt) by exerting force under repetitive motions to remove the unwanted medium. Many cleaning tasks can be categorized as scrubbing, e. g. scrubbing the oven, scrubbing a pan, or scrubbing the floor.

Grind

Grinding is often used in manufacturing such as planing wood. The medium is separated from the surface and is often a waste product. The tool alignment is crucial for the result. Note that grinding of a surface is considered as negligible microscopic penetration here.

Decompose

Decomposing splits the medium into smaller particles. More iterations can possibly lead to smaller particles. Pestle with a mortar is one example.

3.3.2 Discussion

Figure 3.4 illustrates the versatility of wiping actions. The arrangement of the classes corresponds to the wiping tasks introduced in Fig. 1.3. For each action, numerous tools might be suitable to achieve the desired goal. As for the main classification it is possible that not all tasks can be described by only one process model. It is also possible that one particular chore is actually a combination of several wiping actions such as mopping, which is a complex combination of *emitting*, *distributing*, *scrubbing*, and *absorbing*.

The topology of a task can influence the parameter range of the executed action. For example, the force required to move a medium is related to the medium size paired with the surface friction, e. g. smaller shards are harder to remove from a carpet than bigger ones. Also, time has a significant effect on some tasks. For example, if a mug is cleaned right after being used, coffee leftovers can be removed with little effort. If a mug is cleaned the next day, the required force will be higher and more iterations are necessary to remove the leftovers. Hard-coding parameters is inappropriate. Instead, a process model has to incorporate a proper parameterization. The task parameters can be efficiently stored in an object-centric prior knowledge-base. In the particular example of wiping tasks, the Cartesian motion, the Cartesian stiffness and the maximal Cartesian force have to be defined w. r. t. the tool-surface-medium tuple.

The sub-categorization of wiping tasks as defined in Fig. 3.4 allows to allocate a particular process models for each wiping task observed in everyday environments. In particular, a symbolic effect verb catalog is derived including *absorbed*, *skimmed*, *collected*, *emitted*, *distributed*, *processed*, *scrubbed*, *ground*, and *decomposed*. These symbolic descriptors are applicable as *predicates* as defined in the *Planning Domain Definition Language (PDDL)* (Ghallab et al. 1998), the standard language for automated planning. They can be utilized to describe preconditions and effects for symbolic planning. This way, PDDL can be used to define the desired state of the medium in an abstract form. Based on this representation, a symbolic planner can schedule the required actions to solve the tasks symbolically. However, the geometric procedure is still unclear. The representations to cope with this issue are presented in the following chapter.

So far, a deeper investigation has been conducted for one concrete class of compliant manipulation in detail, namely wiping tasks. As discussed in Sect. 3.2.3, the taxonomy may have possible influence on the selection of appropriate control strategies. To verify this, a more detailed analysis for each branch of the tree should be conducted. A possible outcome could be a *standard controller*, respective, a distinct *set of standard parameters* for each branch of the proposed classification. These general purpose strategies could be utilized for previously unseen problems if no specialized control strategy is known to the robot. Moreover, a distinct set of parameters for each branch of the tree could be exploited to automate the parameterization process for new actions, new objects, or new environments. Eventually, a complete set of actions covering each aspect of compliant manipulation is desirable, yet even an incomplete set is already valuable. For example, developing generic action templates for each wiping task according to Fig. 3.4 can already be utilized as basis for common cleaning tasks, which cover almost half of the tasks in domestic environments (Cakmak and Takayama 2013).

3.4 Summary

This chapter introduced a taxonomy for compliant manipulation tasks by investigating the physical contact situation occurring in various domains. Rather than defining a taxonomy on purely geometric features, it was proposed to develop a more abstract taxonomy that integrates the semantic nature of physical contact situations. In addition, a detailed classification of wiping tasks was derived based on the relations observed for the tool-medium-surface tuple, which is fundamental to every wiping tasks. The developed two-step classification constitutes the starting point for the investigations in the remainder of this manuscript. In the following chapter it shall be utilized to derive the necessary representations to plan everyday manipulation tasks.

References

Leidner, Daniel, Alexander Dietrich, Michael Beetz, and Alin Albu-Schäffer. 2016b. Knowledge-Enabled Parameterization of Whole-body Control Strategies For Compliant Service Robots. *Autonomous Robots (AURO): Special Issue on Whole-Body Control of Contacts and Dynamics for Humanoid Robots* 40(3): 519–536.

Leidner, Daniel, Christoph Borst, Alexander Dietrich, and Alin Albu-Schäffer. 2015a. Classifying Compliant Manipulation Tasks for Automated Planning in Robotics. In *Proceedings of the IEEE/RSJ International Conference on Intelligent Robots and Systems (IROS)*, 1769–1776.

Wörgötter, Florentin, Eren Erdal Aksoy, Norbert Kruger, Justus Piater, Ales Ude, and Minija Tamosiunaite. 2013. A Simple Ontology of Manipulation Actions Based on Hand-object Relations. *IEEE Transactions on Autonomous Mental Development* 5(2): 117–134.

Bloomfield, Aaron, Yu Deng, Jeff Wampler, Pascale Rondot, Dina Harth, Mary McManus, and Norman Badler. 2003. A Taxonomy and Comparison of Haptic Actions for Disassembly Tasks. In *Proceedings of the Virtual Reality Conference*, 225–231.

Vukobratović, Miomir K., and Veljko Potkonjak. 1999. Dynamics of Contact Tasks in Robotics. Part I: General Model of Robot Interacting with Environment. *Mechanism and Machine Theory* 34(6): 923–942.

Cutkosky, Mark R. 1989. On Grasp Choice, Grasp Models, and the Design of Hands for Manufacturing Tasks. *IEEE Transactions on Robotics and Automation* 5(3): 269–279.

Bullock, Ian M., Raymond R., Ma, and Aaron M., Dollar. 2013. A Hand-centric Classification of Human and Robot Dexterous Manipulation. *IEEE Transactions on Haptics* 6(2): 129–144.

Daniel Morrow, J., and Pradeep K., Khosla. 1997. Manipulation Task Primitives for Composing Robot Skills. In *Proceedings of the IEEE International Conference on Robotics and Automation (ICRA)*, vol. 4, 3354–3359. IEEE.

Kapandji, Ibrahim Adalbert, and Louis Henri Honoré. 1970. *The Physiology of the Joints: Annotated Diagrams of the Mechanics of the Human Joints*, vol. 1. E. & S. Livingstone London.

Feix, Thomas, Roland Pawlik, Heinz-Bodo Schmiedmayer, Javier Romero, and Danica Kragic. 2009. A Comprehensive Grasp Taxonomy. In *Robotics, Science and Systems: Workshop on Understanding the Human Hand for Advancing Robotic Manipulation*, 2–3.

Stansfield, Sharon A. 1991. Robotic Grasping of Unknown Objects: A Knowledge-Based Approach. *The International Journal of Robotics Research* 10(4): 314–326.

Grebenstein, Markus, Alin Albu-Schäffer, Thomas Bahls, Maxime Chalon, Oliver Eiberger, Werner Friedl, Robin Gruber, Sami Haddadin, Ulrich Hagn, and Robert Haslinger, et al. 2011. The DLR Hand arm System. In *Proceedings of the IEEE International Conference on Robotics and Automation (ICRA)*, 3175–3182.

Chalon, Maxime, Alexander Dietrich, and Markus Grebenstein. 2014. The Thumb of the Anthropomorphic Awiwi Hand: From Concept To Evaluation. *International Journal of Humanoid Robotics (IJHR)* 11: 1450019.

Liu, Jia, Fangxiaoyu Feng, Yuzuko C., Nakamura, and Nancy S., Pollard. 2016. Annotating Everyday Grasps in Action. In *Dance Notations and Robot Motion*, 263–282. Springer.

Ghallab, Malik, Dana Nau, and Paolo Traverso. 2004. *Automated Planning: Theory and Practice*. Morgan Kaufmann.

Williams, David, and Oussama Khatib. 1993. The Virtual Linkage: A Model for Internal Forces in Multi-grasp Manipulation. In *Proceedings of the IEEE International Conference on Robotics and Automation (ICRA)*, 1025–1030.

Leidner, Daniel, Selma Music, and Armin Wedler. 2015b. Robotic Deployment of Extraterrestrial Seismic Networks. In *Proceedings of the 13th Symposium on Advanced Space Technologies in Robotics and Automation (ASTRA)*.

Wimböck, Thomas, Benjamin Jahn, and Gerd Hirzinger. 2011. Synergy Level Impedance Control for Multifingered Hands. In *Proceedings of the IEEE/RSJ International Conference on Intelligent Robots and Systems (IROS)*, 973–979.

Stemmer, Andreas, Alin Albu-Schäffer, and Gerd Hirzinger. 2007. An Analytical Method for the Planning of Robust Assembly Tasks of Complex Shaped Planar Parts. In *Proceedings of the IEEE International Conference on Robotics and Automation (ICRA)*, 317–323.

Hogan, Neville. 1987. Stable Execution of Contact Tasks Using Impedance Control. In *Proceedings of the IEEE International Conference on Robotics and Automation*, vol. 4, 1047–1054.

Xie, Yu, Dong Sun, Chong Liu, Ho Y., Tse, and Shuk H., Cheng. 2010. A Force Control Approach to a Robot-assisted Cell Microinjection System. *International Journal of Robotics Research* 29(9): 1222–1232.

Dietrich, Alexander, Thomas Wimböck, Alin Albu-Schäffer, and Gerd Hirzinger. 2012. Reactive Whole-body Control: Dynamic Mobile Manipulation Using a Large Number of Actuated Degrees of Freedom. *IEEE Robotics & Automation Magazine* 19 (2): 20–33.

Cakmak, Maya, and Leila Takayama. 2013. Towards A Comprehensive Chore List for Domestic Robots. In *Proceedings of the ACM/IEEE International Conference on Human-robot Interaction (HRI)*, 93–94.

Ghallab, Malik, Adele Howe, Dave Christianson, Drew McDermott, Ashwin Ram, Manuela Veloso, Daniel Weld, and David Wilkins. 1998. PDDL - The Planning Domain Definition Language. *AIPS98 Planning Committee* 78(4): 1–27.

Chapter 4
Representations for Compliant Manipulation

Mental representations of objects, actions, motions and effects are fundamental to many cognitive capabilities of humans. These four types of representations influence each other significantly. The interaction between the different representation-levels is already illustrated in Fig. 1.6. In fact, there is evidence that humans maintain the knowledge about actions w. r. t. particular objects (Hebb 2005) and that actions are represented in the human mind by their outcome (Hommel 2009). For example, the typical action one would connect with a broom is sweeping the floor. The desired outcome of sweeping the floor is thereby the accumulation of dust and dirt in order to dispose it afterwards. This reasoning seems natural for humans. However, it is not straightforward for robots to make this connection. This chapter shall therefore develop suitable representations for robotic manipulation that incorporate this reasoning. Accordingly, this chapter is fundamental for the remainder of this manuscript as it is emphasized in Fig. 4.1.

This chapter provides the necessary representations to solve compliant manipulation tasks by means of autonomous robot manipulation. It proposes an equivalent to human-like mental representations, which is one of the keys toward cognitive robots. The representations are introduced in a top-down fashion based on the discussion conducted in Chap. 3. First, a concept to represent task knowledge by means of hierarchical objects is introduced in Sect. 4.1. Based on this, an action representation integrating symbolic and geometric process models is described in Sect. 4.2. A representation to describe the effect on the medium is presented in Sect. 4.3. Eventually, an effect-oriented representation to describe wiping motions is developed in Sect. 4.4.

The proposed representations are based on the findings published in Leidner et al. (2012), Leidner et al. (2014), Leidner et al. (2016b), Leidner et al. (2016a), and Leidner and Beetz (2016). The hierarchical object model as well as Action Templates are introduced in Leidner et al. (2012) and refined afterwards in Leidner et al. (2014)

D. S. Leidner, *Cognitive Reasoning for Compliant Robot Manipulation*, Springer Tracts in Advanced Robotics 127,
https://doi.org/10.1007/978-3-030-04858-7_4

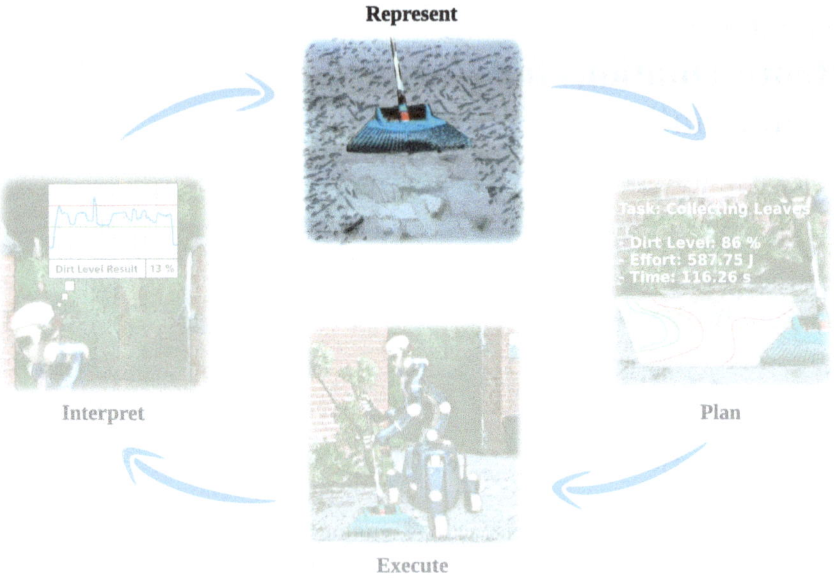

Fig. 4.1 Representing intelligent physical compliance

and Leidner et al. (2016b). The models to describe wiping motions and effects in a goal-oriented manner have been introduced in Leidner et al. (2016a) and later utilized in Leidner and Beetz (2016) to infer the effect of real-world wiping motions.

4.1 Representing Task Knowledge as Object Properties

The representation of object knowledge forms the basis of the reasoning methods to be introduced in this work. In a nutshell, an object-centric approach is introduced to allow a robot to query every kind of information from an object. This may include the fundamental physical properties such as the volumetric model or mass parameters. Moreover, each object is equipped with a set of available handling instructions tailored to the needs of automated robotic manipulation. This object knowledge underpins the planning, execution, and interpretation of compliant manipulation tasks. This mindset is related to the concept of *affordances* introduced by Gibson (1986). Affordances can be described as the set of meaningful actions that an object *affords* to an observer, i. e. a human, an animal, or maybe even a robot.[1] The following section will derive such a data structure based on the well known approach of object-orientation.

As humanoid service robots have to interact with a wide variety of objects, it is necessary to store their respective knowledge in a scalable manner, while maintaining

[1] An overview of affordances applied to robotic manipulation is given by Şahin et al. (2007).

a flexible way of accessing the data. One possible solution to this issue is provided by the *object-orientation* paradigm. As the name suggests, object-orientation is an artificial construct that imitates the characteristic of real world objects. Consequently, describing real world objects in an object-oriented manner is a natural process. The idea behind this approach is the arrangement of information in an hierarchical manner. That is, each entity is represented by an abstract "object", or object *class*. Objects describe properties for an entity, including parameters or functions. Special cases of a generic object class can be derived, where some properties may be inherited, while others are redefined. For example, following the object-oriented programming scheme one could define a generic class of tools and derive a window wiper, which is a special tool to wipe windows. Object classes describe blueprints that are utilized to create concrete object *instances*. Depending on the instantiated object, it is automatically chosen which implementation of a particular function should be executed. For example, wiping a window with a window wiper is different from wiping a window with a sponge. However, one does not have to select between two specific wiping actions when handling either of the objects. Instead, the appropriate action is automatically selected.

Objects are categorized by functionality and hierarchically arranged in the object-oriented paradigm. Each object class defines its particular object properties and available actions (see Sect. 4.2). On the top most level, a generic `_object` class describes universally applicable properties that are inherited by all other sub-classes, such as the fact that every object is described by a unique name. The `_object` class is of abstract nature, i. e. it is not instantiable during run-time, denoted by the leading underscore. In the hierarchy levels below, the main classes of objects found in everyday activities are derived such as, `_structures`, `_tools` or `_containers`. Physical objects may be derived from these abstract classes and inherit their properties. Object classes are later on instantiated to populate the robots internal world model.

Action definitions are stored w. r. t. abstract object classes, while concrete parameters are stored w. r. t. physical objects. The concept behind this idea is that objects of the same class, typically share the way they are handled, while each object exhibits its own parameter range. A concrete example of this procedure is outlined in Fig. 4.2. The `_window_wiper` class provides the information about the `skim` action for all kinds of window wipers. This includes what the goal of the action is symbolically, as well as how it should be performed based on geometric principles, i. e. by means of distinct robot operations. The parameters for these operations are thereby defined by the particular type of window wiper that is used to perform the action. In the example at hand, the `kaercher_wv50_wiper` is used. It contributes with certain parameters such as the wiper with, the force one should apply, and the tool frame which is typically located at the center of the wiper blade. In combination with the information that is provided by a particular window (i. e. the window size and the region of interest), a robot is able to calculate the remaining parameters to perform the task. The concrete implementation of actions is detailed in Sect. 4.2.

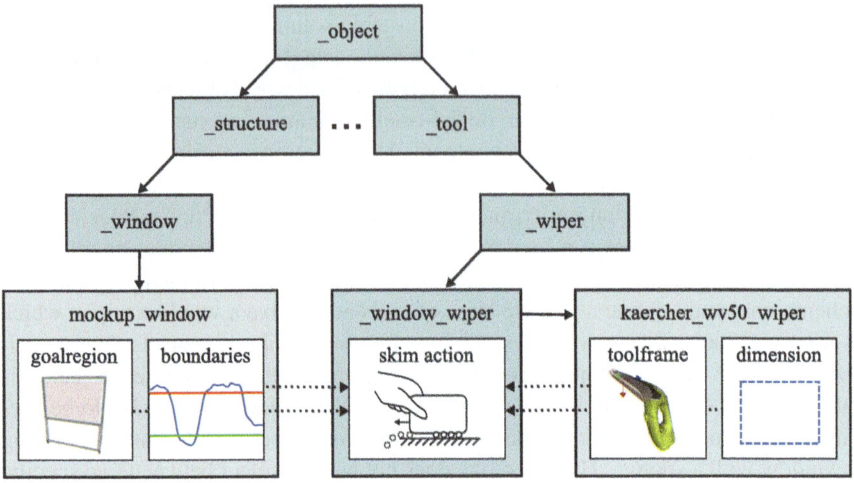

Fig. 4.2 Example ontology for the _window_wiper object class. The skim action is populated w. r. t. the properties of the kaercher_wv50_wiper and the properties of the mockup_window

4.2 Object-Centric Action Representation

Action definitions in robotics and AI differ vastly in the level of abstraction. In AI reasoning, action definitions are usually limited to the discrete description of preconditions and postconditions. In automated planning, several domain specific languages have been developed to describe the nature of tasks and actions. One of them is PDDL introduced in Sect. 2.4, which is the de facto standard language for automated planning. The single elements that are required to specify the action are listed as `parameters`. A list of `preconditions` define the properties that have to hold such that the action is executable. Similarly, `effects` list the nominal changes to the world state. While this descriptive representation is sufficient to describe the outcome of the action, it is not specified how the action has to be performed by the robot. A PDDL definition to describe a car polishing action is listed in Listing 4.1 as an example for a real world scenario.

Listing 4.1 Exemplary PDDL action definition.

```
(:action polish:
    :parameters (?p - polishing_device ?c - car ?
       m - manipulator)
    :precondition (and (picked ?p ?m))
    :effect (and (polished ?c ?m))
)
```

In industrial settings, robot actions are usually programmed by means of continuous motion primitives, such as linear motions in Cartesian space (`LIN`) and

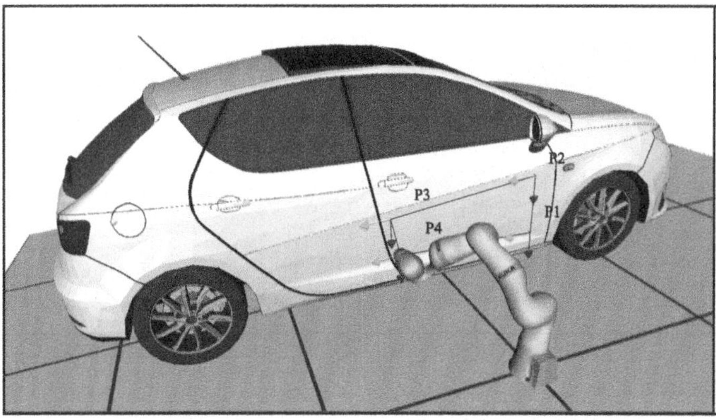


```
polishing.krl

INI

PTP START

LIN P1
LIN P2
LIN P3
LIN P4

PTP HOME

END
```

Fig. 4.3 Classical program for a polishing action with a KUKA robot

joint-space motions to preprogrammed joint positions (`PTP`). Additionally, a tool may be commanded to operate, such as a gripper that opens or closes to grasp or un-grasp an object. The parameters of the single operations are thereby predefined by an expert user. As an example, an industrial polishing robot follows preprogrammed trajectories and activates or deactivates the polishing machine at predefined positions. A set of preprogrammed Cartesian positions can be processed to generate the motions for such an action as shown in Fig. 4.3. In the example at hand, the wiping motion is defined as the joint motion toward a certain `START` position, followed by four linear Cartesian motions along the four points `P1` to `P4` to traverse along the curvature of the car. Eventually, the robot returns to the `HOME` position and the task is finalized. This program describes one way to polish a car with minimal effort and in the shortest possible time. In robotics research, motion planning algorithms are utilized to plan collision-free joint-space motions w. r. t. a geometric world model. The robot is defined based on a volumetric model of its links, given the forward kinematics for the joints (Craig 2005). Inverse kinematics are used to convert Cartesian end-effector positions into joint configurations (Buss 2004). Compared to industrial setups, this is an advancement that enables a robot to cope with previously unseen environments.

Both, the symbolic as well as the geometric action descriptions have major drawbacks that make it impossible to describe robotic manipulation tasks to their full extend. On one hand, the symbolic description written in PDDL is only a descriptive representation of the action. The polish action is merely a narration for the actual procedure. It does not include any information on how the action should be performed. Instead, it outlines the constraints of the action symbolically. The only precondition stated is that the polishing device needs to be picked by the manipulator. The resulting effect is described by the word "polished". While humans are able to interpret this definition based on experience and observations, a robot would be unable to execute the task solely based on this information. On the other hand, the industrial specification for the same task is restricted to the particular example. The motion of

Action Template: _object.pick

I Symbolic Representation

```
'''
(:action _object.pick:
    :parameters (?o - _object ?m - _manipulator ?s - _surface)
    :precondition (and(free ?m) (on ?o ?s))
    :effect (and(bound ?o ?m) (not(free ?m)) (not(on ?o ?s)))
)
'''
```

II Geometric Representation

```
def pick(self, manipulator, surface):
    ''' approach, grasp, and lift an object from a surface '''

    graspset = odb.get_property(self.type, 'graspset', manipulator)
    for grasp_candidate in graspset:
        if grasp_candidate in self.history:
            continue
        self.history.append(candidate)
        grasp = grasp_candidate
        break

        if grasp is None:
            raise RuntimeError('no more alternatives')
        else:
            self.grasp = grasp

    operations = [
        ('move_hand', manipulator, g.approach_grasp),
        ('plan_to_frame', manipulator, g.approach_frame, self.frame),
        ('plan_to_frame', manipulator, g.grasp_frame, self.frame),
        ('bind', manipulator, self.name),
        ('move_hand', manipulator, g.pre_grasp),
        ('move_hand', manipulator, g.grasp),
        ('move_straight', manipulator, txyz(0, 0, 0.1))
    ]
    return operations
```

Fig. 4.4 The action template for a pick action provided by the generic `_object` class

the robot is defined by means of constant values that happen to match the contour of the car. A robot that executes this motion with a polishing device attached to the end-effector may be able to solve the task of polishing a certain car. However, this information is not explicitly stated and the action cannot be generalized to other cars or objects. Consequently, the polishing effect can be considered a side-effect of the robot motions. To truly foster intelligent behavior, a robot has to be aware of both perspectives as it is incorporated by the concept of *Action Templates*.

4.2.1 Action Templates

Action Templates constitute the representation developed for this work to describe robot actions based on symbolic predicates, as well as w. r. t. geometric process models. This dualism is necessary to enable a robot to understand the meaning of an action on one hand, and to realize the action by means of goal oriented tool motions on the other hand. In general, Action Templates describe the symbolic header of an action based on PDDL syntax, and the geometric procedure with executable Python code. An example Action Template for a simple pick action is visualized in Fig. 4.4. In this example, the action describes the process of grasping an object to lift it from a surface. The Action Template for this procedure is provided in a single file, stored w. r. t. the definition of the generic `_object` class. That is, every object that does not specify otherwise, can be picked up from a target surface by a robot that has at least one manipulator available. From a symbolic perspective, this seems to be a valid assumption. However, it has to be assessed if this action is really feasible. Figure 4.4 illustrates this two layers of abstraction as two separate blocks. The reasoning procedure is interleaved as it will be detailed in Chap. 5. In the following, the segments are introduced individually.

The first segment is the symbolic header, repeated in Listing 4.2. This part of the code is realized as a comment section within a Python file. The symbolic header holds the PDDL description for the action. That is, it describes the symbolic preconditions and the nominal effects for the action. The parameters are defined based on the object classes that are available in the object storage. In the example at hand, the object of interest is of type `_object`, and can thus be any object that derives from the object hierarchy outlined in Fig. 4.2. A `_manipulator` is required to grasp the object from a `_surface`, which is the super class for any object with one or more planar supporting planes, e. g. tables, shelves, or trays. The preconditions that have to hold are that the object has to be located on a surface, and that the manipulator is free (i. e. not holding something). The effects are that the object is bound to the manipulator, the manipulator is not free anymore, and the object no longer touches the surface.

Listing 4.2 Pick action template: Semantic header.

```
1 '''
2 (:action _object.pick:
3     :parameters (?o - _object ?m - _manipulator
              ?s - _surface)
4     :precondition (and(free ?m) (on ?o ?s))
5     :effect (and(bound ?o ?m) (not(free ?m))
               (not(on ?o ?s)))
6 )
7 '''
```

The second element of the Action Template describes the geometric process model by means of executable Python code (see Listing 4.3). The code implements a function that incorporates the same parameters as the PDDL definition (cf. line 3 in

Listing 4.2). All object parameters (denoted by the `self` pointer) are available within the scope of this function. Moreover, the function can access the properties of all additional objects that participate in the action (i. e. the robot `manipulator` and the `surface` object). Consequently, the `_object.pick` action is described based on the properties stored w. r. t. the object that is grasped, the manipulator performing the action, and the object that serves as the target surface. This information is the key to ground the symbolic meaning of the Action Template to a geometric process model.

Listing 4.3 Pick action template: Geometric body.

```
def pick(self, manipulator, surface):
    ''' approach, grasp, and lift an object from
        a surface '''
```

Each Action Template is structured in a similar fashion. Initially, geometric properties required to implement the process model have to be gathered from the object storage. In the example at hand, the grasp set of the object is accessed as listed in Listing 4.4. The for loop in the code segment indicates that multiple grasps are available (cf. line 2). In fact, almost every manipulation task can be solved in several different ways. For example, multiple grasp sets are suitable to pick up an object from a surface. The alternative grasp sets include the hand transformation and the finger configuration to approach the object collision-free, as well as the final hand transformation and the finger configuration to grasp the object. The reasoning mechanism to select the most appropriate alternative is detailed in Chap. 5.

Listing 4.4 Pick action template: Knowledge-based definition of geometric parameters.

```
1      graspset = odb.get_property(self.type,
           'graspset', manipulator)
2      for grasp_candidate in graspset:
3          if grasp_candidate in self.history:
4              continue
5          self.history.append(candidate)
6          g = grasp_candidate
7          break
8
9      if g is None:
10         raise RuntimeError('no more alternatives')
11     else:
12         self.grasp = g
```

The gathered information is eventually utilized to instrument a list of operations that is tailored to result in the desired outcome. An elemental operation may describe a primitive motion in the Cartesian object space, a joint motion for the end-effector, or the realization of a particular controller setting. In the example at hand, a sequence of robot motions has to be generated in order to reach the object, grasp it, and lift it from the table as it is done in Listing 4.5. The operations are implemented by means of robot specific modules. For example, the `plan_to_frame` operation implements a joint motion for robots with traditional manipulators (e. g. Rollin' Justin). In contrast, the same operation may be implemented as a sequence of flight maneuvers for a drone. This modular design is discussed in Chap. 5.

Listing 4.5 Pick action template: List of robot operations.

```
    operations = [
        ('move_hand', manipulator,
            g.approach_grasp),
        ('plan_to_frame', manipulator,
            g.approach_frame, self.frame),
        ('plan_to_frame', manipulator,
            g.grasp_frame, self.frame),
        ('bind', manipulator, self.name),
        ('move_hand', manipulator, g.pre_grasp),
        ('move_hand', manipulator, g.grasp),
        ('move_straight', manipulator,
            txyz(0, 0, 0.1))
    ]
    return operations
```

In conclusion, Action Templates provide the means to develop complex robot skills in a generic form. While Action Templates are in principle independent of specific robot capabilities, it is possible to integrate robot specific reasoning methods by means of external modules. As a result, one particular skill can be made available for arbitrary robots with different capabilities.

4.2.2 Action Templates for Compliant Manipulation

Picking up an objects is a well suited example to illustrate the basic concept of Action Templates. However, defining an Action Template is not always straightforward, as it is not always possible to map a particular symbolic predicate to a numerical outcome. For example, pick-and-place tasks can be described w. r. t. the relocation of an object, which can be represented by the translation and rotation of its transformation matrix. Only a few simple primitive operations are required to describe the pick-up sequence for an object. Most of the time it is thereby not even necessary to grasp the tool in a meaningful way, i. e. not to solve a certain problem. For example, to only relocate an object, it is not necessary to grasp a mug at the handle. This is only needed if, for example, a person wants to drink from it. It is sufficient to select any grasp that is reachable to pickup the object. The concrete parameterization of the action has a minor role.

In contrast, compliant manipulation tasks often require a much richer understanding of the problem. Especially wiping actions are often ambiguous w. r. t. their purpose. The same wiping motion may produce very different effects depending on *how*

Action Template: _window_wiper.skim

I Symbolic Representation

```
'''
(:action _window_wiper.skim:
    :parameters (?t - _tool ?m - _medium
                 ?s - _surface ?a1 - _manipulator)
    :precondition (and (picked ?t ?a1)
                       (applied ?s ?m))
    :effect (and (not (applied ?s ?m)))
)
'''
```

II Geometric Representation

```
def skim(self, medium, surface, manipulator):
    ''' skim the medium off the surface '''

    trajectory = self.get_task_trajectory(target.dimension)
    initial_frame = trajectory[0]
    initial_config = robot.get_configuration()

    operations = [
        ("plan_to_frame", manipulator, tool.grasp_frame, initial_frame),
        ("set_stiffness", manipulator, tool.tcp, tool.stiffness),
        ("set_force", manipulator, tool.tcp, tool.force),
        ("follow_task_motion", manipulator, trajectory, tool.grasp_frame),
        ("plan_to_config", initial_config),
    ]
    return operations
```

Fig. 4.5 The action template for a skimming action as it is provided by the abstract `_window_wiper` class

it is executed exactly. Depending on how much force is applied, a wiping motions may dry a wet surface or remove sticky dirt. When moving the wiper into contact with the target surface, the robot has to react with compliant behavior to adapt the tool correctly. The compliance parameters can therefore be defined as a property of the tool as depicted in Fig. 4.5. In particular, the concept of *object impedance* (Wimböck 2013) allows to define virtual potential fields in the coordinates of the *Tool Center Point (TCP)* of an object. For example, the desired contact force, as well as the desired Cartesian stiffness can be defined w. r. t. the TCP of the window wiper which is located in the center of the wiper blade. The controller of the robotic manipulator reacts thereby compliantly, such that the tool aligns w. r. t. the curvature of the target surface. The integration of compliance parameters as part of Action Templates will be detailed in Chap. 6.

The same reasoning applies for the effect of wiping motions. In comparison to the pick-and-place tasks, it becomes obvious that the relocation of the tool (e. g. a sponge or a window wiper) is not the desired effect of wiping actions. In fact, the

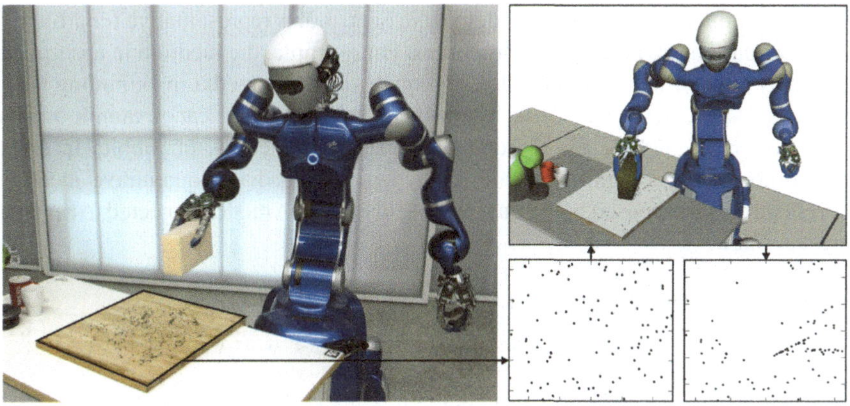

Fig. 4.6 Change estimation of the particle distribution in contact with a sponge during wiping. A uniform particle distribution is assumed in the example at hand. Upon contact with the sponge, the particle model is updated w. r. t. the properties of the medium. Here, solid particles are simulated to be pushed by the sponge

classification of wiping tasks discussed in Sect. 3.3 reveals that the main goal of wiping tasks is to manipulate a medium. The successful execution of wiping tasks requires a substantial amount of geometric reasoning as the purpose of wiping might be the collection, the distribution, the absorption of particles or any of the other tasks introduced in Fig. 3.4. In order to perform the commanded tasks successfully the robot has to carefully select the task parameters in a continuous parameter space. These aspects have to be incorporated within a representation of wiping actions, such that a robot to is able to reason about individual task parameterization. In order to incorporate this within the concept of Action Templates, an Action Template can refer to more complex code elements such as planning algorithms for goal-oriented wiping motions as it is detailed in the following sections.

4.3 Representing the Effects of Wiping Actions

According to the classification of compliant manipulation tasks conducted in Chap. 3, compliant manipulation tasks shall be represented w. r. t. the semantic contact situation between the manipulated objects and the environment. This high-level of abstraction allows to define semantic actions in accordance with the descriptive manner of the action definitions for automated planning (Ghallab et al. 2004). To further derive geometric process models, wiping actions have to be geometrically represented based on the relation between the tool, the medium to be manipulated (e. g. particles or liquids), and the target surface. Consequently, service robots need an abstract representation of wiping effects in order to reason about wiping actions in a generalized form.

The medium in wiping tasks, as defined in Sect. 3.3, is representative for arbitrary liquids or particles with different properties. For example, the medium in absorption tasks may be a variation of dust, dirt, or liquids. Possible media in skimming tasks would be water, detergent, or snow on the windshield of a car. Example media for collect tasks include the shards of a broken mug, leaves, or rubble. In order to incorporate these significantly different types of media, a qualitative medium representation is proposed based on a particle distribution model projected to a planar target surface

$$\begin{aligned}\mathcal{P} = \big\{(x_1, y_1), (x_2, y_2), \ldots, (x_{\mathrm{N}}, y_{\mathrm{N}}) \mid x_i, y_i \in \mathbb{R} \wedge \\ x_{\min} \leq x_i \leq x_{\max} \wedge y_{\min} \leq y_i \leq y_{\max}\big\},\end{aligned} \tag{4.1}$$

where N particles (x_i, y_i) are distributed within the boundaries of the target surface $(x_{\min}, x_{\max}, y_{\min}, y_{\max})$. A random particle distribution is shown in Fig. 4.6. The scenario shows a kitchen environment where the humanoid robot Rollin' Justin is instructed to collect bread crumbs distributed on a chopping board. In some scenarios, the medium can be perceived visually if the particles are big enough or the liquid is not transparent. However, especially water and other transparent liquids or dust and other small particles are very hard to perceive in camera images. Consequently, the real distribution of the medium cannot be modeled by the robot. In this case, a uniform distribution is assumed as it is shown for the initial particle distribution in Fig. 4.6.

The main purpose of the proposed particle representation is to have a naive predictive model of the effects of wiping actions. The applied tool-particle contact model considers the exact CAD data of the tool and the position of each particle. Depending on the type of the tool and the properties of the medium (i. e. solid particles or liquids), the contact results in different effects. For example, if a sponge is simulated to wipe a liquid, the resulting effect is the absorption of the liquid, which is implemented as a delete operation of the respective particles. In case of a solid medium, the contact with the sponge pushes the particles in parallel to the direction of the tool motion. This contact behavior is utilized to simulate the collect and the skim actions. The different effects on the manipulated medium are visualized in Fig. 4.7.

Automated planning solutions applied for AI-based reasoning typically describe desired *state changes* $\mathcal{S}_{t_0} \rightarrow \mathcal{S}_{\mathrm{g}}$ as semantic postconditions (i.e. effects) in a descriptive manner. Accordingly, a cognition-enabled robotic systems has to be able to ground abstract high-level commands to geometric procedures. That is, it has to interpret human-comprehensible task definitions such as `(collected breadcrumbs chopping_board)`, and generate the respective tool motions. Therefore, the proposed particle distribution model allows to define abstract goals by means of changes to the particle distribution $\mathcal{P}_{t_0} \rightarrow \mathcal{P}_{\mathrm{g}}$. In particular, it is possible to describe the desired state change based on constraint definitions for the particle set. Based on this description, a geometric planner can be developed to generate goal-oriented tool motions. The constraint definitions for all identified wiping actions are defined as follows.

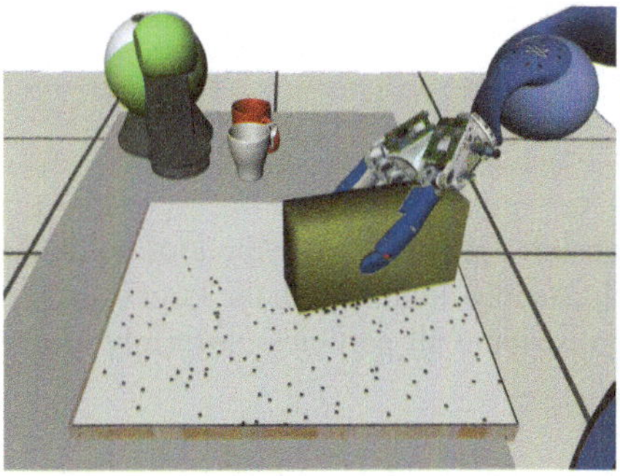

(a) Pushing solids.

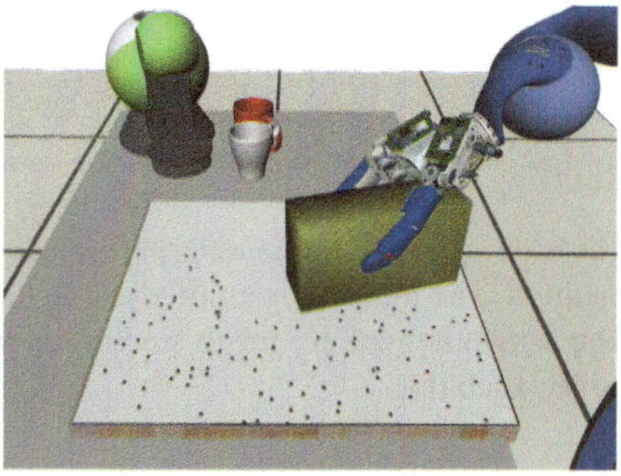

(b) Deleting liquids.

Fig. 4.7 The particle model to represent effects of wiping motions. On the left, the particles are pushed such that they accumulate in front of the sponge. On the right, the particles are deleted as they touch the sponge

Absorbing

The absorb action is the first action to be investigated in this work. It occurs e. g. in vacuuming, dusting, or soaking up water with a sponge. The particle effect (i. e. the absorption) is independent of the tool motion, as long as contact is made with the entire region of interest. The particular effect of the action is that all particles are bound to the tool and vice versa removed from the target surface, such that

$$\mathcal{P}_{\mathrm{g,absorb}} = \emptyset\,. \tag{4.2}$$

Collecting

The second removal action is to collect the medium, where the particles are pushed upon contact with the tool. The goal state $\mathcal{S}_{\mathrm{g}}$ is geometrically represented by a dedicated goal node n_{goal} on the target surface. This goal region is limited to the maximum extent of the tool, represented by the diameter d_{s}. The effect is an accumulated pile of particles, representable as

$$\begin{aligned}\mathcal{P}_{\mathrm{g,collect}} = \big\{&(x_1, y_1), (x_2, y_2), \ldots, (x_{\mathrm{N}}, y_{\mathrm{N}}) \,|\, x_i, y_i \in \mathbb{R} \wedge \\ &||(n_{\mathrm{goal},x}, n_{\mathrm{goal},y}) - (x_i, y_i)|| \leq d_{\mathrm{s}}/2\big\}\,.\end{aligned} \tag{4.3}$$

Skimming

Skimming is related to collecting the medium. Upon contact with the tool the particles are pushed away. The semantic goal $\mathcal{S}_{\mathrm{g}}$ for this action can be described as a geometric state where all particles are located outside the boundaries of the target surface. Depending on the task, only a subset of boundaries of the target surface may be used to specify the valid goal region, where

$$\mathcal{P}_{\text{g,skim}} = \{(x_1, y_1), (x_2, y_2), \ldots, (x_N, y_N) \mid x_i, y_i \in \mathbb{R} \wedge (x_{\min} > x_i \vee x_i > x_{\max} \vee y_{\min} > y_i \vee y_i > y_{\max})\}. \tag{4.4}$$

Emitting

The emit action is the counterpart to the absorb action. Initially the tool holds the medium. The medium is emitted as the tool gets into contact with the target surface. The semantic goal $\mathcal{S}_g$ for this action is to distribute all particles uniformly on the target, such that

$$\mathcal{P}_{\text{g,emit}} = \{(x_1, y_1), (x_2, y_2), \ldots, (x_N, y_N) \mid x_i, y_i \in \mathbb{R} \wedge x_i \in \mathcal{U}(x_{\min}, x_{\max}) \wedge y_i \in \mathcal{U}(y_{\min}, y_{\max})\}. \tag{4.5}$$

Distributing

The goal function $\mathcal{S}_g$ for the distribution of particles is the same as the goal of the emit action. However, the initial state is different as the particles are already located on the target in this case. The goal is described by the uniform distribution $\mathcal{U}$ of the particles in both dimensions of the target surface

$$\mathcal{P}_{\text{g,distribute}} = \{(x_1, y_1), (x_2, y_2), \ldots, (x_N, y_N) \mid x_i, y_i \in \mathbb{R} \wedge x_i \in \mathcal{U}(x_{\min}, x_{\max}) \wedge y_i \in \mathcal{U}(y_{\min}, y_{\max})\}. \tag{4.6}$$

Processing

Processing describes an approach to distribute a medium w.r.t. a complex goal function, defined by a task specific and possibly inhomogeneous target area $G_r(x, y)$. In the simplest case, it can be described as a subset of the collect action. However, it might be that certain directions to reach the goal location are non-permissible due to geometric constraints. The goal function $\mathcal{S}_g$ can be grounded as

$$\mathcal{P}_{\text{g,process}} = \{(x_1, y_1), (x_2, y_2), \ldots, (x_N, y_N) \mid x_i, y_i \in \mathbb{R} \wedge x_i, y_i \in G_r(x, y)\}. \tag{4.7}$$

Scrubbing

Scrubbing is categorized as one of three modification actions (cf. Fig. 3.4). That is, the properties of the medium are actively altered by the influence of a tool. Accordingly, the particle model cannot describe the goal state by means of a certain goal location. Instead, more abstract goals are defined.[2]

In case of scrubbing, one medium of type a (initially hold by the tool) influences a second medium of type b in such a way, that it is not anymore adhesive to the target

[2]Abstract goals cannot accurately be described by means of geometric constraints of the particle distribution. Especially the description for the decomposition action (4.10) is therefore not anymore conform with the standards of set-builder notation.

surface. Therefore, the matching constraint definition describes a state in which each particle of type a is exclusively bound to a particle of type b.

$$\mathcal{P}_{\text{g,scrub}} = \{(x_{a,1}, y_{a,1}), \ldots, (x_{a,\text{N}}, y_{a,\text{N}}), (x_{b,1}, y_{b,1}), \ldots, (x_{b,\text{M}}, y_{b,\text{M}}) \mid x_{a,i}, y_{a,i} \in \mathbb{R} \wedge x_{b,j}, y_{b,j} \in \mathbb{R} \wedge (x_{a,i}, y_{a,i}) = (x_{b,j}, y_{b,j})\} . \tag{4.8}$$

Grinding

The grinding action is typically found in manufacturing. It describes the procedure to remove parts of the surface by repetitive motions with a tool. Accordingly, the medium originates from the surface and it is less relevant in terms of the semantic goal $\mathcal{S}_{\text{g}}$. Yet, the skillful motion of a tool may eventually result in a state where the medium is uniformly degraded, such that

$$\mathcal{P}_{\text{g,grind}} = \{(x_1, y_1), (x_2, y_2), \ldots, (x_{\text{N}}, y_{\text{N}}) \mid x_i, y_i \in \mathbb{R} \wedge x_i \in \mathcal{U}(x_{\min}, x_{\max}) \wedge y_i \in \mathcal{U}(y_{\min}, y_{\max})\} . \tag{4.9}$$

Decomposing

The decomposition of the medium is the attempt to divide the medium into its sub-components by applying pressure. It can be semantically described by a state $\mathcal{S}_{\text{g}}$ in which each particle is subdivided into two smaller particles a and b, respectively, where

$$\mathcal{P}_{\text{g,decompose}} = \{(x_1, y_1), (x_2, y_2), \ldots, (x_{\text{N}}, y_{\text{N}}) \mid x_i, y_i \in \mathbb{R} \wedge (x_i, y_i) \rightarrow (x_{a,i}, y_{a,i}), (x_{b,i}, y_{b,i})\} . \tag{4.10}$$

These constraint definitions are the basis to qualitatively reason about the effect of wiping actions and rate the performance accordingly. Throughout the remainder of this manuscript the three removal actions absorbing, skimming, and collecting are investigated w. r. t. the developed particle model. It is shown how this qualitative representation can be applied for simulation based prediction, as well as the evaluation of previously executed wiping motions recorded by means of episodic memories. Therefore, position information and force information is combined in simulation in order to account for real world effects as detailed in Chap. 7. This allows to update the effect prediction as well as the anticipated performance estimation. Eventually, the robot is able to question its own motions and reschedule additional wiping actions to increase the performance subsequently. The following sections describe how wiping motions are represented that match the goal definitions described above.

4.4 Representing Wiping Motions

A general definition for the act of wiping has been developed in Chap. 1. It defines wiping motions as motions where *a tool is slid along a target surface in contact by following task-oriented Cartesian workspace trajectories. During contact, task-specific forces are applied along the surface normal while a certain stiffness is adjusted to align with the surface curvature.*. Accordingly, wiping motions have to be represented by means of a tool trajectory and an appropriate set of compliance parameters. The latter one can be realized by means of object properties as described in Sect. 4.1. A desired contact force vector can be defined w. r. t. the TCP of the tool. Ideally, the stiffness values can be designed such that the tool aligns with the curvature of the target surface. However, the tool trajectory depends on the dimensions of the target surface, the distribution of the medium, and the desired goal state, i. e. the particular wiping task. It is not trivial to define goal-oriented wiping motions in a generalized form.

The geometric principle behind wiping motions can be considered as a *coverage path planning problem* (Gabriely and Rimon 2001). The wiping motions have to cover the entire region of interest for a target surface (e. g. a table surface), given a certain distribution for a medium to be manipulated (e. g. dirt). Based on this, the removal actions to be investigated can be formulated w. r. t. the coverage of a target surface within the bounds of the proposed particle distribution model. For example, if a robot shall be commanded to remove spilled coffee from a table, it will have to cover the entire table in the worst case, but at least the regions with spilled coffee. Furthermore, each action has to satisfy the anticipated semantic goal state. In the example at hand, the absorption of the coffee would be the main goal. In this work, such problems are represented by means of *Semantic Directed Graphs (SDGs)*

$$\mathbf{SDG} = f(\mathcal{P}, \mathcal{S}_{\mathrm{g}}, \mathcal{G}_{t_0}), \tag{4.11}$$

which are specified as a function of the particle set $\mathcal{P}$, the semantic goal state $\mathcal{S}_{\mathrm{g}}$, and the initial geometric state of the environment $\mathcal{G}_{t_0}$. Altogether, SDGs project a graph structure on a planar surface, where

- each node n_i represents a waypoint (i. e. the position and orientation) for the Cartesian tool motion w. r. t. the TCP of the handled object,
- the edge (n_i, n_{i+1}) connecting two nodes represents the interpolated tool motion in contact with the surface.

This section shall only introduce the representation of SDGs. The concrete implementation and the generation of the graphs is detailed in Sect. 5.3. To give an outline, three particular example SDGs are visualized in Fig. 4.8. The graph nodes (cyan boxes) are initially distributed w. r. t. the target surface and the initial particle distribution $\mathcal{P}$ (black dots), creating the basis to plan different wiping actions. Semantically meaningful connections (blue lines) are made in order to achieve the desired effect, i. e. absorbing, skimming, or collecting particles represented by the green boxes

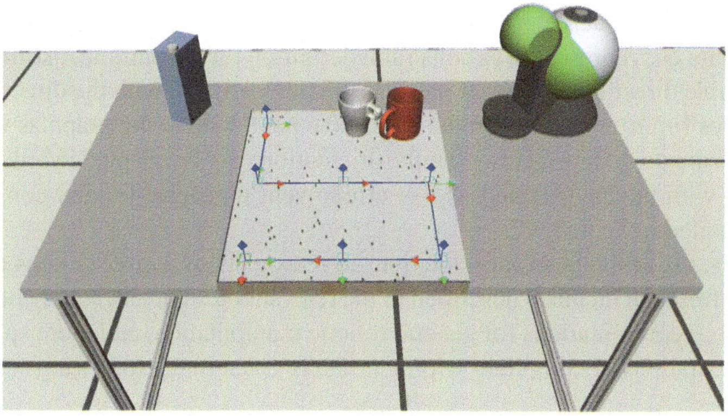

(a) Absorbing

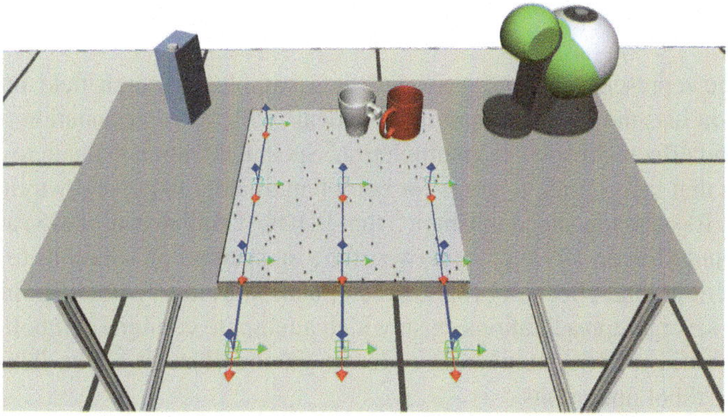

(b) Skimming

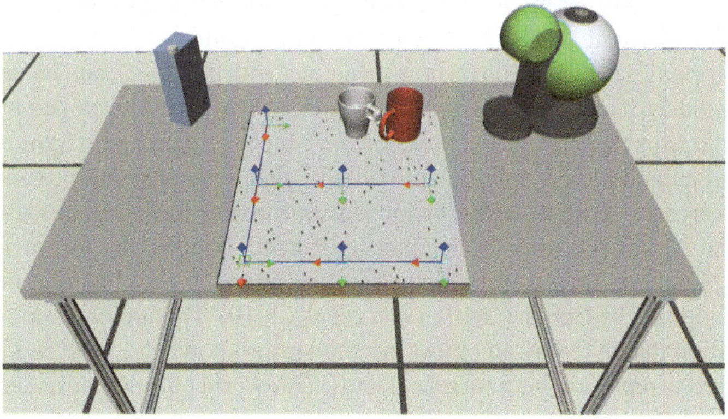

(c) Collecting

Fig. 4.8 Example of SDGs for the three removal actions. The nodes n_i are visualized by cyan boxes. The edges n_i, n_{i+1} are shown as blue lines. The goal positions are highlighted as green boxes (see Sect. 5.3.1). To achieve this, each SDG exhibits its own strategy to concur with the goal state $\mathcal{S}_g$, represented as constraint definition for the particle distribution $\mathcal{P}_g$. The principle direction of the tool motion is thereby aligned with the direction of the graph edges (n_i, n_{i+1}), i. e. pointing toward the next node in the graph as visualized by the coordinate systems. However, the orientation may be adapted in order to avoid collisions with the environment $\mathcal{G}_{t_0}$ or circumvent unreachable joint configuration as it is detailed in Sect. 5.3.2.

Notably, the developed representations to reason about wiping actions differ significantly from the methods described in the relevant literature. A survey on the state of the art on representations for general robotic manipulation, and more specifically, for representations of compliant wiping tasks, is given in the following section.

4.5 Related Work

Knowledge representation and reasoning is a complex research field on its own. This section lists the relevant literature that deals with the representation of objects, actions, and effects for robotic manipulation. Special attention is given to the representations that incorporate features for compliant manipulation and wiping tasks in particular. It is already stated in Chap. 2 that PDDL (Ghallab et al. 1998) and Prolog (Sterling and Shapiro 1994) present two of the most common solutions to represent objects, actions, and effects for robotic manipulation from a symbolic point of view. However, several representations exist, which attempt to combine symbolic features with geometric implementations in order to ground abstract action definitions to executable robot operations.

Several authors apply object-centric reasoning to define robot manipulation skills. Kallmann and Thalmann (1998) propose to store articulation trajectories within the scope of objects in order model the interaction between human agents and so-called *Smart Objects* in virtual worlds. Among others the authors propose to store object intrinsic properties, information on how to interact with the object, and object-specific behavior models. This approach is derived from the methods developed by Levison (1996), who proposes to use the well known object-oriented paradigm to classify objects and augment the symbolic domain with hierarchical properties and actions. These actions are populated with concrete data at run time. Based on this, as symbolic planner can be utilized to plan a sequence of agent actions. In case of failure the symbolic planner is forced to re-plan. Another approach for modeling object data has been proposed by Belkin (2010); Gheta et al. (2010). The authors propose a world representation that is fed by an object-oriented prior knowledge base and the robots sensor inputs to represent internal robot states. This world representation serves as the belief state of the robot, which is utilized as information hub for the sub-components of the robot control program.

The concept of affordances has been widely adopted as design principle of robot actions. A survey on the topic was given by Şahin et al. (2007). Dalibard et al. (2010) introduced the concept of *Documented Objects*, i. e. objects that provide manipulation information to a robotic agent. In particular, Documented Objects describe articulation trajectories for moveable objects such as doors and drawers. An inverse kinematics approach is utilized to translate this information into robot joint motions. A similar approach based on geometric action representations to teleoperate a robot is introduced by Hart et al. (2015). The authors propose to use *Affordance Templates* to command robot actions on-line. The approach is based on templates in form of CAD models that provide common interaction possibilities, such as pushing a button, opening a lid, or turning a valve. This approach was utilized for the DARPA Robotics Challenge. There is no symbolic action representation provided as the templates are designed for general purpose use. A related approach is given with *Object-Action Complexes (OACs)* introduced by Krüger et al. (2011). OACs describe a symbolic representations of continuous sensorimotor experience. They are defined as a triple consisting of an execution specification, an effect prediction function, and a statistical success measure. OACs can be applied to various problems ranging from low-level reactive behavior to high-level deliberative planning with application to robotic manipulation.

Another possible way to define robot manipulation task is the design of constraint based motion descriptions. A formal method suitable to describe actions in the context of for automated planning was introduced by Bartels et al. (2013). They describe a constraint-based approach for the representation of robot actions. The deep integration of this representation with the robot control framework allows to specify geometric interaction on a semantic level. As an example, the authors describe how a robot can be commanded to flip a pancake with a spatula. A similar approach is presented with the *iTaSC* framework. Vanthienen et al. (2013) describe how this framework can be used to describe the force-interaction procedure during table wiping task.

A recently introduced data-driven manipulation procedure is concerned with the use of the Internet as knowledge resource for robots (Tenorth et al. 2011). On one hand, the authors try to interpret informations from do-it-yourself web pages for humans (Beetz et al. 2011). On the other hand they develop a world wide web for robots called *KnowRob* (Tenorth and Beetz 2009; Waibel et al. 2011; Tenorth et al. 2012). A robot can download information about its environment, the included objects, as well as complete action recipes to execute a given task rather than solve it from scratch. Symbolic connections are linked by using the ontology web language *OWL* (McGuinness et al. 2004).

Kunze et al. (2011) suggest to represent robot actions by means of event calculus (Kowalski and Sergot 1989). In particular it is shown how the interaction of a robot in a physics simulation can be logged and translated into symbolic time-interval-based first-order representations. Based on this representation it is possible to describe events based on narratives that are grounded to numerical values. Furthermore, Kunze et al. introduce an approach to reason about the semantic effect of tools interacting with the environment based on naive process models. The authors propose to utilize dynamic simulations to envision the effects of robot actions. As part of this simulation, a particle model is introduced that reacts with certain behavior as the particles get in contact with objects and tools. In one particular example, the effect of a sponge contacting liquids is simulated, which results in the absorption of the liquid. A more abstract representation is provided by Pablo Lana et al. (2015). They represent robotic manipulation tasks in an algebraic form, which incorporates poses, velocities and forces. The work is of particular interest as the authors evaluate their approach in a simulated window cleaning experiment.

4.6 Summary

The mental representation of objects, actions, motions, and their effects is one of the keys toward human-like cognitive reasoning capabilities. Humans use these representations for problem solving and the prediction of task outcomes. In this chapter, an equivalent of these representations was developed for robotic agents. The basis was given with the hierarchical representation of functional object classes. The object classes provide the knowledge that is required to represent actions and their effects, respectively. In particular, Action Templates were introduced to describe actions from an object-centric perspective, ground symbolic action descriptions to geometric process models, and parameterize primitive robot operations. Semantic Directed Graphs (SDGs) were introduced to represent wiping motions based on a goal-oriented graph model. In addition, a particle model was introduced to represent the effect of wiping motions on a qualitative level. As a result, wiping motions can be directly represented in the effect-space. In summary, the introduced representations are fundamental for the reasoning methods developed in the remainder of this manuscript.

References

Bartels, Georg, Kresse, Ingo, and Beetz, Michael. 2013. Constraint-Based Movement Representation Grounded in Geometric Features. In *Proceedings of the IEEE/RAS International Conference on Humanoid Robots (ICHR)*, 547–554.

Beetz, Michael, Klank, Ulrich, Kresse, Ingo, Maldonado, Alexis, Mösenlechner, Lorenz, Pangercic, Dejan, Rühr, Thomas, and Tenorth, Moritz. 2011. Robotic Roommates Making Pancakes. In *Proceedings of the International Conference on Humanoid Robots (ICHR)*, 529–536.

Belkin, Andrey. 2010. Object-Oriented World Modelling for Autonomous Systems. In *Proceedings of the Joint Workshop of Fraunhofer IOSB and Institute for Anthropomatics*, 231. KIT Scientific Publishing.

Buss, Samuel R. 2004. Introduction to Inverse Kinematics with Jacobian Transpose, Pseudoinverse and Damped Least Squares Methods. *IEEE Journal of Robotics and Automation* 17 (1–19): 16.

Craig, John J. 2005. *Introduction to Robotics: Mechanics and Control*, vol. 3. Pearson Prentice Hall Upper Saddle River.

Dalibard, Sébastien, Nakhaei, Alireza, Lamiraux, Florent, and Laumond, Jean-Paul. 2010. Manipulation of Documented Objects by a Walking Humanoid Robot. In *2010 10th IEEE-RAS International Conference on Humanoid Robots (Humanoids)*, 518–523, IEEE.

Gabriely, Yoav, and Elon Rimon. 2001. Spanning-Tree based Coverage of Continuous Areas by a Mobile Robot. *Annals of Mathematics and Artificial Intelligence* 31 (1–4): 77–98.

Ghallab, Malik, Howe, Adele, Christianson, Dave, McDermott, Drew, Ram, Ashwin, Veloso, Manuela, Weld, Daniel, and Wilkins, David. 1998. PDDL-The Planning Domain Definition Language. *AIPS98 Planning Committee*, 78(4): 1–27.

Ghallab, Malik, Dana, Nau, and Paolo, Traverso. 2004. *Automated Planning: Theory and Practice*. Morgan Kaufmann.

Gheta, Ioana, Heizmann, Michael, Belkin, Andrey, and Beyerer, Jürgen. 2010. World Modeling for Autonomous Systems. In *Proceedings of the Advances in Artificial Intelligence (KI)*, 176.

Gibson, James J. 1986. *The Ecological Approach to Visual Perception*, Routledge.

Hart, Stephen, Dinh, Paul, and Hambuchen, Kimberly. 2015. The Affordance Template ROS Package for Robot Task Programming. In *Proceedings of the IEEE International Conference on Robotics and Automation (ICRA)*, 6227–6234.

Hebb, Donald Olding. 2005. *The Organization of Behavior: A Neuropsychological Theory*, Psychology Press.

Hommel, Bernhard. 2009. Action Control According to TEC (Theory of Event Coding). *Psychological Research PRPF* 73 (4): 512–526.

Kallmann, Marcelo, and Daniel Thalmann. 1998. Modeling Objects for Interaction Tasks. In *Proceedings of the Eurographics Workshop on Computer Animation and Simulation*, vol. 98, 73–86.

Kowalski, Robert and Sergot, Marek. 1989. A Logic-based Calculus of Events. In *Foundations of Knowledge Base Management*, 23–55. Springer.

Krüger, Norbert, Christopher Geib, Justus Piater, Ronald Petrick, Mark Steedman, Florentin Wörgötter, Aleš Ude, Tamim Asfour, Dirk Kraft, Damir Omrčen, et al. 2011. Object-Action Complexes: Grounded Abstractions of Sensory-Motor Processes. *Robotics and Autonomous Systems* 59 (10): 740–757.

Kunze, Lars, Dolha, Mihai Emanuel, Guzman, Emitza, and Beetz, Michael. 2011. Simulation-based Temporal Projection of Everyday Robot Object Manipulation. In *Proceedings of the International Conference on Autonomous Agents and Multiagent Systems (AAMAS)*, 107–114.

Leidner, Daniel, Borst, Christoph, and Hirzinger, Gerd. 2012. Things Are Made for What they are: Solving Manipulation Tasks by Using Functional Object Classes. In *Proceedings of the IEEE/RAS International Conference on Humanoid Robots (ICHR)*, 429–435.

Leidner, Daniel, Dietrich, Alexander, Schmidt, Florian, Borst, Christoph, and Albu-Schäffer, Alin. 2014. Object-Centered Hybrid Reasoning for Whole-Body Mobile Manipulation. In *Proceedings of the IEEE International Conference on Robotics and Automation (ICRA)*, 1828–1835.

Leidner, Daniel, and Michael, Beetz. 2016. Inferring the Effects of Wiping Motions based on Haptic Perception. In *Proceedings of the IEEE/RAS International Conference on Humanoid Robots (ICHR)*, 461–468.

Leidner, Daniel, Bejjani, Wissam, Albu-Schäffer, Alin, and Beetz, Michael. 2016a. Robotic Agents Representing, Reasoning, and Executing Wiping Tasks for Daily Household Chores. In *Proceedings of the International Conference on Autonomous Agents and Multiagent Systems (AAMAS)*, 1006–1014.

Leidner, Daniel, Dietrich, Alexander, Beetz, Michael, and Albu-Schäffer, Alin. 2016b. Knowledge-enabled Parameterization of Whole-Body Control Strategies for Compliant Service Robots. *Autonomous Robots (AURO): Special Issue on Whole-Body Control of Contacts and Dynamics for Humanoid Robots*, 40 (3): 519–536.

Levison, Libby. 1996. *Connecting Planning and Acting via Object-Specific Reasoning*. Ph.D. thesis, University of Pennsylvania.

McGuinness, Deborah L, Van Harmelen, Frank, et al. 2004. OWL Web Ontology Language Overview. *W3C Recommendation*, 10 (10): 2004.

Pablo Lana, Ernesto, Vilhena Adorno, Bruno, and Andrey Maia, Carlos. 2015. A New Algebraic Approach for the Description of Robotic Manipulation Tasks. In *Proceedings of the IEEE International Conference on Robotics and Automation (ICRA)*, 3083–3088.

Şahin, Erol, Çakmak, Maya, Doğar, Mehmet R, Uğur, Emre, and Üçoluk, Göktürk. 2007. To Afford or not to Afford: A New Formalization of Affordances Toward Affordance-based Robot Control. *Adaptive Behavior*, 15 (4): 447–472.

Sterling, Leon, and Ehud Y, Shapiro. 1994. *The Art of Prolog: Advanced Programming Techniques*. MIT press.

Tenorth, Moritz, and Michael, Beetz. 2009. Knowrob-knowledge Processing for Autonomous Personal Robots. In *Proceedings of the IEEE/RSJ International Conference on Intelligent Robots and Systems (IROS)*, 4261–4266, IEEE.

Tenorth, Moritz, Ulrich Klank, Dejan Pangercic, and Michael Beetz. 2011. Web-enabled Robots. *IEEE Robotics & Automation Magazine* 18 (2): 58–68.

Tenorth, Moritz, Perzylo, Alexander, Lafrenz, Reinhard, and Beetz, Michael. 2012. The RoboEarth Language: Representing and Exchanging Knowledge about Actions, Objects, and Environments. In *Proceedings of the IEEE International Conference on Robotics and Automation (ICRA)*, 1284–1289.

Vanthienen, Dominick, Robyns, Steven, Aertbeliën, Erwin and De Schutter, Joris. 2013. Force-sensorless Robot Force Control within the Instantaneous Task Specification and Estimation (iTaSC) Framework. In *Benelux Meeting on Systems and Control*.

Waibel, Markus, Michael Beetz, Javier Civera, RRaffaello D'Andrea, Jos Elfring, Dorian Galvez-Lopez, Kai Häussermann, Rob Janssen, JMM Montiel, Alexander Perzylo, Björn Schießle, Moritz Tenorth, Oliver Zweigle, and MJG van de Molengraft. 2011. RoboEarth. In *IEEE Robotics & Automation Magazine*, 18 (2): 69–82.

Wimböck, Thomas. 2013. Controllers for Compliant Two-Handed Dexterous Manipulation. Ph.D. thesis, Vienna University of Technology.

Chapter 5
Planning Everyday Manipulation Tasks

This chapter constitutes the second of the four main chapters in the concept of Intelligent Physical Compliance as depicted in Fig. 5.1. It addresses the question on how symbolic task descriptions can be translated into meaningful robot operations in the context of wiping tasks. To achieve this, three main elements are investigated, namely, symbol grounding and planning with symbols, planning algorithms for mobile manipulation, and effect-space planning for wiping tasks. The aspects of symbol grounding and semantic planning are discussed in Sect. 5.1. The reasoning about mobility is covered in Sect. 5.2. Section 5.3 shows how wiping motions can be planned based on the representations introduced in Sects. 4.3 and 4.4, and how this representation can be used to predict the performance of the action qualitatively.

The findings presented in this chapter are based on the following articles. The hybrid reasoning approach is introduced in (Leidner et al. 2012), later refined in (Leidner et al. 2016b), and extended toward mobile manipulation in (Leidner and Borst 2013; Leidner et al. 2014b). The concept of effect-space planning for wiping tasks was mainly developed in (Bejjani 2015; Leidner et al. 2016a).

5.1 Symbol Grounding and Semantic Planning

The ability to plan the execution of manipulation tasks based on mental simulations is one of the keys to master the various problem instances that humans encounter in their everyday life. According to the concept of cognitive control (Norman and Shallice 1980), conscious problem solving is especially conducted for new situations and failure recovery. A human will generate a plan to solve a given task by subdividing the problem into elemental actions and operations. These actions and operations are mentally processed and the outcome is compared to the desired goal state. For humans, this reasoning is done effortlessly. However, this issue poses a

D. S. Leidner, *Cognitive Reasoning for Compliant Robot Manipulation*, Springer Tracts in Advanced Robotics 127,
https://doi.org/10.1007/978-3-030-04858-7_5

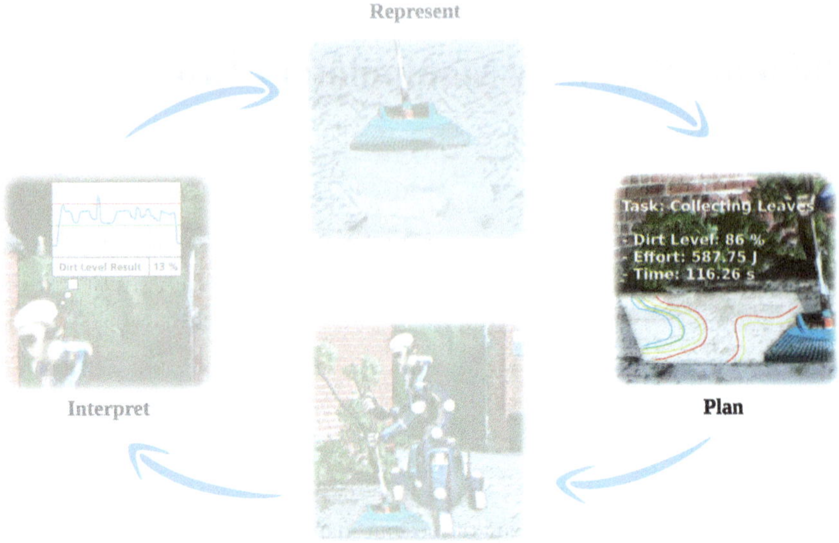

Fig. 5.1 Planning intelligent physical compliance

challenging problem for robots. In particular, a robot has to reason about the desired goal state from a symbolic point of view and translate it into a geometric equivalent. This abstraction is not straightforward and requires concrete process models that match the symbolic outcome. Moreover, a robot may have to concatenate a set of actions to reach a goal. Furthermore, it has to incorporate possible failure states arising from deviations on the geometric level. For example, even though the symbolic precondition for a pouring action is to hold a bottle, it becomes geometrically unfeasible to pour from a bottle if the bottle is grasped at the outlet. An alternative grasp has to be selected in order to execute the task successfully. Depending on the task, it might even be necessary to choose an entirely different sequence of actions to solve the problem. This problem is commonly referred to as *symbol grounding* (Harnad 1990).

The symbol grounding problem raises the question on how purely symbolic mental representations can become meaningful, by connecting them to objects, actions and effects in the real world. In that sense, it is the question how symbols get their meaning. In cognitive science, this is often considered a problem that is very hard to solve in general as it is for example argued by the *Chinese Room Argument*[1] (Searle 2001). In the particular case of robotics, symbol grounding refers either to the problem of generating motor commands for robotic manipulators from AI-level action representations (i.e. *action grounding*), or to the problem of representing ef-

[1]The Chinese Room Argument argues that a person would be able to answer written Chinese questions, given all answers by means of Chinese symbols, without knowing the meaning of the symbols.

fects of robotic manipulation by means of abstract symbols (i.e. *effect grounding*). Both problems are not trivial and discussed in this work. The problem of action grounding is discussed within this section. The problem of effect grounding is investigated in Chap. 7.

Creating artificial cognitive intelligence is an attempt to create human-like intelligence. As a part of this, symbol grounding becomes one of the biggest challenges that has to be solved, before universal service robots can be deployed in everyday environments. One particular issue is thereby the lack of a generic representation to describe actions in terms of symbolic descriptions, as well as w. r. t. geometric process models. With the introduction of Action Templates, this work proposes one possible way toward such a generic representation. However, it is important to integrate this representation into the greater picture. Action Templates are therefore established as the fundamental building block to take part in semantic planning. Semantic planning, as it is described here, incorporates symbolic planning on one hand, and geometric planning on the other hand. One possibility to tackle this issue is combined task and motion planning, sometimes referred to as *hybrid planning*. The following section proposes a hybrid planning system based on object-centered reasoning.

5.1.1 Object-Centered Hybrid Reasoning

This section introduces the concept of hybrid reasoning as an attempt to solve arbitrary robotic manipulation tasks by means of object-centered action descriptions, i.e. Action Templates. Task requirements vary w. r. t. the objects that participate in the task execution, the environmental conditions, and the robot that executes the actions. This demands detailed knowledge about the task and the involved objects. Therefore, it was proposed in Sect. 4.1 to categorize objects in a hierarchical structure w. r. t. their functionality and additionally store process models to define arbitrary manipulation instructions. This information is utilized here to reason about robotic manipulation tasks. The architecture is outlined in Fig. 5.2.

The object storage (upper left) provides prior knowledge for all available objects. The objects are hierarchically arranged and categorized by functionality as introduced in Sect. 4.1. Objects of the same class share the same process models to handle them and can therefore be manipulated in the same manner, but under consideration of their specific properties such as size and shape. The world representation (lower left) provides the current state for the environment of the robot. Objects as described in the object storage are instantiated with specific symbolic and geometric properties. Action Templates are utilized for hybrid planning (marked blue). As introduced in Sect. 4.2, Action Templates consist of two segments. The first segment provides symbolic action definitions for symbolic planning in the PDDL language (upper right). The second segment specifies geometric instructions as executable code to implement the symbolic effects by the use of geometric reasoning modules such as navigation, motion planning or dynamics computations. Individual robot components are thereby addressed to solve the commanded task on the geometric level (lower

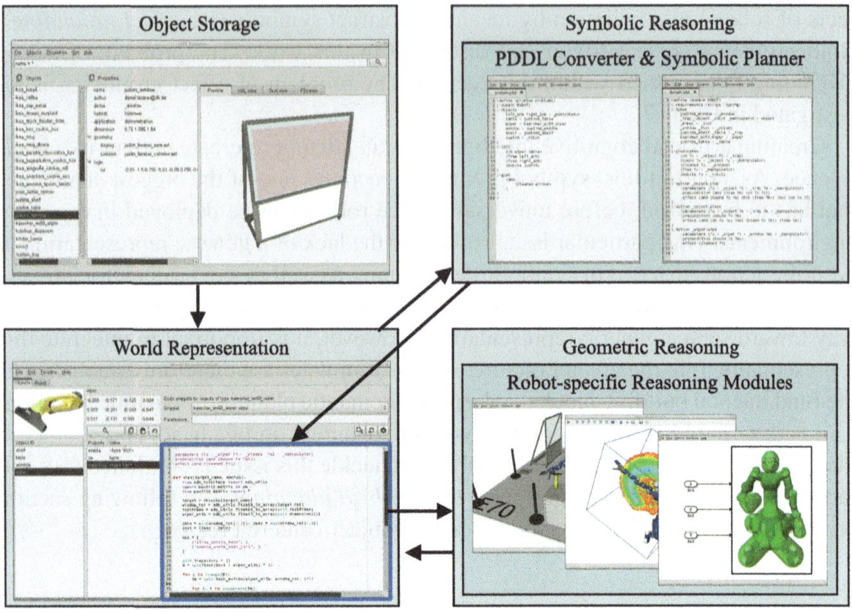

Fig. 5.2 The flow chart for solving manipulation tasks within the object context

right). Action Templates constitute the main element for the approach on object-centered hybrid reasoning in a two-step approach:

First, the current world state is parsed in order to generate a symbolic domain description. Therefore, all objects that are currently instantiated are queried for the actions they afford. This information is gathered from the symbolic header of the Action Templates. Additionally, the symbolic properties are collected from the knowledge base. With this information, and the additional information of the object types, it is possible to describe a fully qualified PDDL domain file and the PDDL problem definition. The symbolic domain is thereby only filled with information of the current state, which limits the search space to a minimum. Based on this, a symbolic planner is able to generate a symbolic transition leading to the desired goal state.

In the second step, the main part of the Action Templates is revisited in order to ground the symbolic action definition to geometric process models. The geometric reasoning procedure is listed in Algorithm 5.1. The Action Templates α_i are sequentially parsed for their operations $\boldsymbol{\Omega}$. This is done for each step of the symbolic transition $\boldsymbol{\mathcal{T}}_{sym}$ that is generated by the symbolic planner. If one geometric reasoning step is successfully simulated, the resulting robot commands $\boldsymbol{\Phi}$ are added ($\boldsymbol{\mathcal{T}}_{geo} \frown \langle \boldsymbol{\Phi} \rangle$) to the geometric transition vector $\boldsymbol{\mathcal{T}}_{geo}$. The next actions are simulated until the entire symbolic transition is processed. If one particular action cannot be simulated in the intended way, it is revisited with an alternative parameter set σ_j. Possible errors include insufficient reachability, collisions with obstacles, or missing task information such as grasp sets for a particular manipulator. The set of available

Algorithm 5.1: Geometric reasoning procedure.

Input: The symbolic transition $\mathcal{T}_{sym}$ generated by the symbolic planner, containing the list of actions to be simulated.

Output: The geometric transition $\mathcal{T}_{geo}$, which contains the list of low-level commands $\boldsymbol{\Phi}$ to be executed by the robot. The outcome is based on the simulations conducted for each action α_i, yielding a list of operations $\boldsymbol{\Omega}$

$\mathcal{T}_{geo} \leftarrow$ List()
while *Length*($\mathcal{T}_{sym}$) $\neq 0$ **do**
    solved $\leftarrow$ False
    $\alpha_i \leftarrow$ GetListEntry($\mathcal{T}_{sym}$, 0)
    for $\sigma_j \in$ *YieldAlternatives*(α_i) **do**
        $\boldsymbol{\Omega} \leftarrow$ ParseActionTemplate(α_i, σ_j)
        $\boldsymbol{\Phi} \leftarrow$ SimulateOperations($\boldsymbol{\Omega}$)
        if $\boldsymbol{\Phi} = \emptyset$ **then**
            continue
        $\mathcal{T}_{geo} {}^\frown \langle \boldsymbol{\Phi} \rangle$
        RemoveListEntry($\mathcal{T}_{sym}, i$)
        solved $\leftarrow$ True
        break
    if *solved* = *False* **then**
        return List()
return $\mathcal{T}_{geo}$

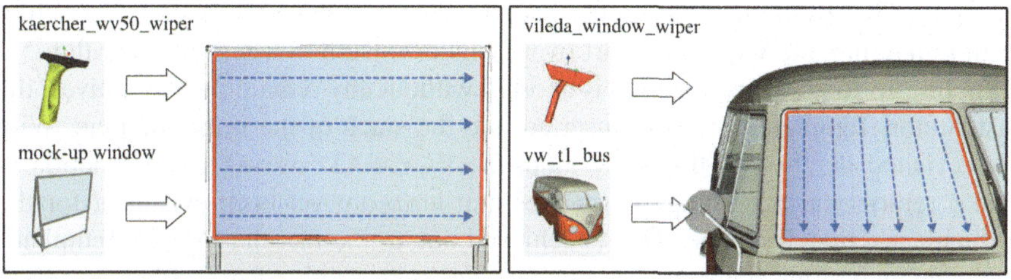

Fig. 5.3 Example parameterization of the geometric process models of the *skim* Action Template with different wipers and windows. The target surface is highlighted in red. The derived tool trajectories are drawn in blue

alternative parameters depends on the particular action. For example, a pick-up action may provide alternative grasp orientations and configurations to grasp an object from different angles. The set of available alternatives to place an object consists of possible locations on the target surface. The modules to simulate the geometric execution are provided by the robot performing the tasks. Accordingly, specific needs and requirements can be incorporated. For example, an unmanned aerial vehicle has to use a different geometric planner for motion generation than a ground-based humanoid. In the case of Rollin' Justin out-off-the-shelf software is used almost exclusively. In particular, the *Fast Downward* planner by Helmert (2006) is used for symbolic planning and OpenRAVE by Diankov (2010) for geometric planning.

Action Templates define process models at the highest possible level of abstraction to provide generic control strategies. Individual variations are defined at the object level, i.e. scrubbing a plate with a squared sponge applies the same Action Template with different parameters (e. g. different trajectory, force, and stiffness). If necessary, it is nonetheless possible to specialize the process model for specialized object types. This concept is illustrated in Fig. 5.3 at the example of skimming a window with a window wiper. Depending on the window pane a different trajectory design might be required. This concept is applied to windows directly, or objects that contain windows such as the bus on the right. The examples in this section are handcrafted for illustration purposes. A more generic solution to plan wiping motions is presented in Sect. 5.3. The trajectory design is stored as part of the window object. It is determined through parameters provided by distinct tools to calculate concrete Cartesian workspace motions for the execution.

5.1.2 Backtracking

The description in the previous section assumes that there exists at least one alternative parameter set for each Action Template that leads to a successful task execution. However, it is quite likely that one step of the geometric reasoning procedure fails to succeed for all of the available alternatives. Should the predefined set of alternatives turn out to be insufficient, geometric backtracking is initiated to find a prior action with remaining alternatives to start over. This extension is listed in Algorithm 5.2 from line 15 to 20. If the for-loop is aborted without any remaining alternatives, the last Action Template α_{i-1} is re-inserted into the stack of the symbolic plan. With the revisited iteration of the while-loop, the `YieldAlternatives` function in line 5 returns the remaining alternatives that have not yet been evaluated for the particular Action Template. This is achieved due to the fact that Action Templates maintain an internal status of their previous parameterization from earlier execution attempts. Accordingly, a new branch in the search tree is expanded each time an Action Template is revisited as it is visualized in Fig. 5.4. This way, the algorithm explores the entire search space, spanned by the sequence of actions matching the symbolic transition. If the provided symbolic transition is completely unfeasible, it is removed from the symbolic domain and the symbolic planner is called once again in order to find an alternative solution on the symbolic level as shown in Fig. 5.5. The figures illustrate the selection of geometric alternatives, the geometric backtracking mechanism, as well as the symbolic backtracking step in the conceptual example of scrubbing a mug.

In the example at hand, the robot is commanded to clean the inner side of a mug with a sponge brush. As described in Sect. 4.1, the available actions to handle the objects are provided by the object storage in the form of Action Templates. In addtion, the object storage holds task relevant information such as grasp configurations and symbolic properties (e. g. the fact that a mug has to be grasped in order to clean it). Both objects have to be picked up first, and a detergent has to be absorbed by the

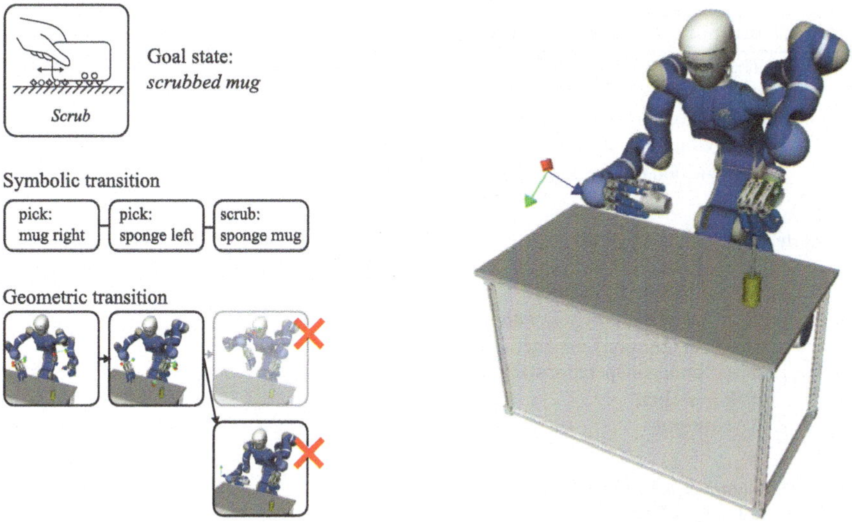

Fig. 5.4 Alternative parameter sets and geometric backtracking are used to find a feasible configuration to clean the inner side of a mug with a sponge. Given the choice made for the first alternative, the hand is covering the mug, which prevents any sucessful task completion

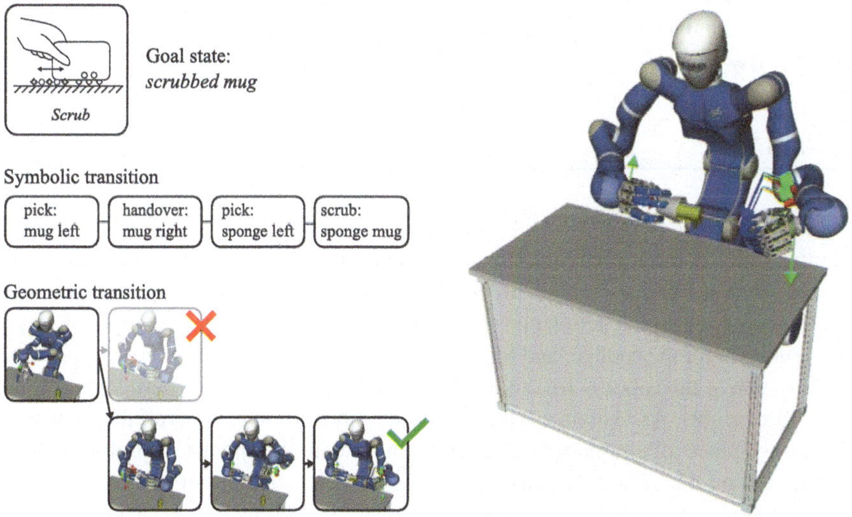

Fig. 5.5 None of the alternatives explored in Fig. 5.4 are sucessful, such that symbolic backtracking has to be initialized. The symbolic planner explores the symbolic state space anew. As a result, the planner adds a handover action such that the inner side of the mug can be cleaned

Algorithm 5.2: Geometric reasoning with integrated geometric backtracking.

Input: The symbolic transition $\mathcal{T}_{sym}$ generated by the symbolic planner, containing the list of actions to be simulated.

Output: The geometric transition $\mathcal{T}_{geo}$, which contains the list of low-level commands $\mathbf{\Phi}$ to be executed by the robot. The outcome is based on the simulations conducted for each action α_i, yielding a list of operations $\mathbf{\Omega}$

$\mathcal{T}_{geo} \leftarrow$ List()
while *Length*($\mathcal{T}_{sym}$) $\neq 0$ **do**
    solved $\leftarrow$ False
    $\alpha_i \leftarrow$ GetListEntry($\mathcal{T}_{sym}$, 0)
    for $\sigma_j \in$ *YieldAlternatives*(α_i) **do**
        $\mathbf{\Omega} \leftarrow$ ParseActionTemplate(α_i, σ_j)
        $\mathbf{\Phi} \leftarrow$ SimulateOperations($\mathbf{\Omega}$)
        if $\mathbf{\Phi} = \emptyset$ **then**
            continue
        $\mathcal{T}_{geo} \frown \langle \mathbf{\Phi} \rangle$
        RemoveListEntry($\mathcal{T}_{sym}$, i)
        solved $\leftarrow$ True
        break
    if *solved = False* **then**
        if *Available*(α_{i-1}) **then**
            InsertListEntry($\mathcal{T}_{sym}$, α_{i-1}, 0)
            RemoveListEntry($\mathcal{T}_{geo}$, $i-1$)
            continue
        else
            return List()
return $\mathcal{T}_{geo}$

sponge. Only if these preconditions apply, can the scrubbing action be scheduled. The anticipated effect is that the dirt is `scrubbed` away from the mug. The available alternatives for the pick action consist of two different grasps for the mug, one grasp configuration for the sponge, and four different orientations to scrub the mug with the sponge, where the robot is free to choose the manipulator for the particular actions (i.e. left or right). As visualized, the top grasp is not suitable to scrub the mug with the sponge since all alternatives to clean the mug with the sponge are colliding with the right hand of the robot. Figure 5.4 visualizes two of the four attempts that are simulated in order to infer the feasibility of the scrubbing action. The algorithm backtracks to the first action with available alternatives, which is the pick action for the mug. However, based on the initial position of the mug (i.e. bottom-down on the table), the robot is unable to pick it up in order to clean the inner side, assuming there is only one top grasp and one bottom grasp available.

As a consequence of insufficient alternatives to grasp the mug from the side, the hybrid reasoning algorithm has to backtrack symbolically. By removing the possibility to directly grasp the mug with the right hand, the symbolic domain is pruned in such a way, that the symbolic planner has to schedule a different sequence of actions. That is, the mug has to be grasped with the left hand first. As the left hand grasps the mug from above, it can be handed over to the right hand in a bottom-down fashion. Afterwards, the left hand can be used to pick the sponge from the table. Eventually, this allows the robot to clean the mug as the inner curvature of the mug is no longer blocked by the right manipulator.

The backtracking mechanism is an exhaustive search algorithm. The reasoning procedure is therefore complete within the bounds of the discrete search space defined by the available alternatives for all scheduled Action Templates. The method is guaranteed to find a solution if one set of alternatives describes a feasible geometric action sequence. Consequently, it can be stated that the more alternatives are available for the individual Action Templates, the higher the probability to find a feasible solution.

5.1.3 Execution

This chapter is mainly focusing on the motion generation aspect for robotic manipulation. The integrated hybrid reasoning procedure plans the motions based on symbolic and geometric goal states. The output of this procedure is a full set of robot commands to perform the task in the real world. This includes all the necessary controller commands. In fact, Action Templates are designed to integrate symbolic parameters, geometric motion parameters, and control parameters. This aspect is detailed in Chap. 6, where the fundamental controller design for compliant manipulation, the parameterization of the control level, and the concrete execution is discussed at the example of several wiping tasks executed with the humanoid robot Rollin' Justin.

5.1.4 Discussion

Symbol grounding is a major challenge for autonomous robot manipulation. Only by reasoning about this issue, can a robot execute meaningful actions based on abstract task specifications. This section presents an approach to solve this issue by using functional object classes. Action Templates are arranged in the object context to describe arbitrary process models. They are used to populate the domain of a symbolic planner and ground the geometric actions to the robot. Within an Action Template, arbitrary sub-components of a robot can be addressed. This concept brings the desired goal and the corresponding action description into focus rather than the capabilities of a robot. Effects to the world are thus applied in a natural way.

5.2 Mobile Manipulation

Mobility is crucial for service robots in human households. Objects in human environments are widely spread over different rooms and different storage positions. Even simple fetch-and-carry tasks require a robot to navigate from one object to another. The main cause for moving toward an object is inssuficient reachability. However, finding a suitable base position to manipulate a certain object is not trivial. Limiting factors are collisions with obstacles, end-effector kinematics, and the reachability of the manipulator.

Especially, compliant manipuilation tasks occupy extended *Regions of Interest (ROIs)*. Having a closer look on compliant manipulation tasks, it becomes evident that these tasks often occupy more workspace than a similar task without compliance. For example, placing an object on a table can be solved without relying on compliant contact since the complete trajectory is known before it is executed. When inserting a peg into a hole, a robot has to execute additional, previously unknown evasion motions to avoid jamming the peg. These motions might require to significantly extend the nominal workspace of the action. It is therefore crucial to optimally place the manipulator such that it does not fall into local minima, nor collide with obstacles, or fail to reach the target configuration. This problem is even more obvious in the running example of wiping tasks. Wiping tasks are often applied to large regions such as window panes, table surfaces, the entire floor of a room, or maybe even the entire area of a park as it was depicted in the introductory example shown in Fig. 1.1. In such cases, it is advantageous to place the robot optimally to maximize the reachability for these regions, such that it does not have to reposition itself unnecessarily over and over again.

An approach to generate robot motions that match a symbolic goal state has been proposed in the previous section. Only motion planning algorithms for robotic manipulators were discussed in the proposed architecture. One could simply extend the motion planner by the DOF of a mobile base to solve mobile manipulation tasks as shown by Schulman et al. (2013). However, if the goal position to manipulate an object is ambiguous, this approach cannot be applied directly. For example, one does not always know where to place a mug on a shelf, before the shelf has been approached. The concrete goal position depends on the space that is blocked by other objects, the reachability of the robotic manipulator, and the desired orientation for the object to be placed. In particular, this extends the goal position, to a goal region. Consequently, it is not possible to define an exact goal pose, nor a specific configuration beforehand. Therefore, a novel approach to solve this issue is discussed in this section. The approach takes advantage of pre-calculated reachability information to find the most adequate base position in terms of reachability for the objects involved in the task execution.

5.2.1 *Reasoning About Reachability*

Manipulating objects with a mobile robot raises the question of finding a suitable position to place the mobile base of a robot, such that task-related objects become reachable. Zacharias et al. (2007) suggest to use *Capability Maps* as external module for symbolic planners to answer this question with geometric background knowledge. Capability Maps describe a discrete representation of the workspace of robotic manipulators. The Capability Map for the right arm of the humanoid robot Rollin' Justin is visualized in Fig. 5.6. The picture illustrates Rollin' Justin in its zero-position, i.e. with all joint values set to be zero. This pose presents the maximal extentension of the robot, naturally matching the border of the Capability Map.

For each of the individual voxels in the workspace of the manipulator of the robot, the *reachability index* is computed

$$r = \frac{R}{N}, \tag{5.1}$$

where N is the maximum number of hypothetically reachable discrete positions in the voxel, and R is the number of positions which are reachable for the manipulator. The color coding of the map illustrates the reachability for each voxel, where dark red

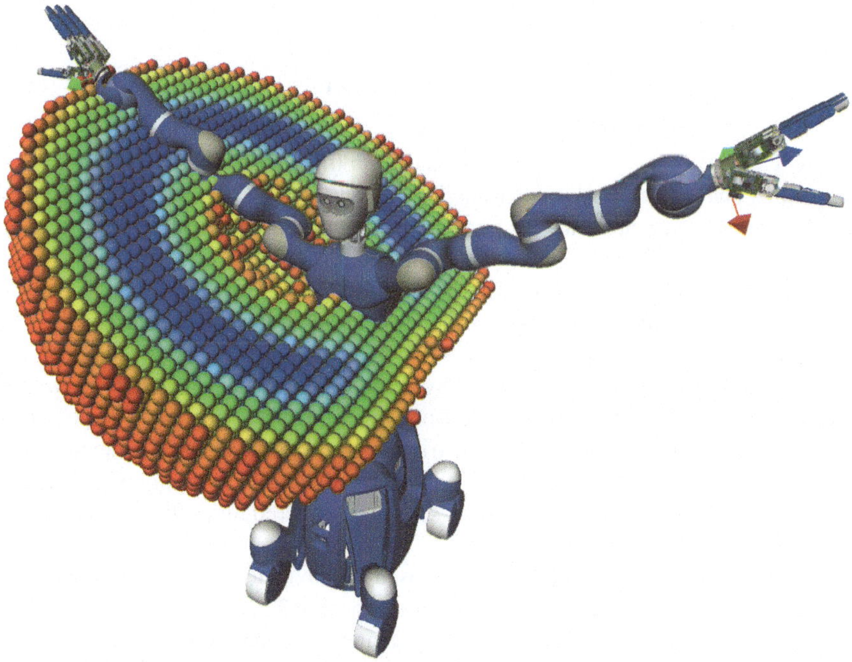

Fig. 5.6 Cross section of the Capability Map for Rollin' Justin's right arm

is unreachable ($r = 0.0$) and dark blue is fully reachable ($r = 1.0$). The maximum reachability index r_{max} for the arms of the humanoid robot Rollin' Justin is 0.833. That means there is no position in the workspace of the robot that is reachable from any possible orientation. This is true for most robotic manipulators. However, the illustration reveals that there is a coherent region in the workspace of the robot that shows particularly good reachability. This coherency can be exploited in the search for optimal robot base positions.

It is shown that reaching for objects (Stulp et al. 2012a) as well as following workspace trajectories (Zacharias et al. 2009) can be guided by the use of capability maps. In particular, Stulp et al. (2012a) calculate feasible positions to reach for an object under consideration of the reachability of the robot. Zacharias et al. (2009) propose to query Capability Maps to optimize the base position of a robot to realize Cartesian task trajectories such as opening a cupboard.

This section expands these approaches to cover whole ROIs in the workspace of the manipulator. The regions are gathered out of the object database and may vary depending on the type of object. Small objects that a robot can manipulate with one manipulator (e.g. mugs and bottles) form a ROI based on the dimension of the object bounding box. Bigger objects define a ROI according to their purpose. For example, tables and shelves define their ROIs based on the dimension of the storage surface. The key is to maximize the mean reachability for these areas, by optimizing the reachability index for the robotic manipulator within the boundaries of the ROI. The mean reachability value r_{roi} is therefore optimized by adapting the base position $\boldsymbol{p}_{\text{base}}$ of the robot and the configuration of the torso $\boldsymbol{q}_{\text{torso}}$,

$$\boldsymbol{x}_{\max} = \arg\max_{\boldsymbol{x} \in \mathcal{D}} \; r_{\text{roi}}(\boldsymbol{x}), \tag{5.2}$$

where $\boldsymbol{x} = (\boldsymbol{p}_{\text{base}}^T, \boldsymbol{q}_{\text{torso}}^T)^T$. The initial seed to optimize the location of the base is defined by the current object position. The joint values for the initial torso configuration are set to be upright. The optimal intersection between the capability map of the right manipulator and the ROI of a window wiper and a window is shown in Fig. 5.7. The ROI for the window wiper is defined by the object bounding box, while the ROI for the window is defined by the area of the window pane as it is stored in the object database

The solution vector $\boldsymbol{x}_{\max}$ is obtained by maximizing the mean reachability within the region of interest in the restricted domain $\mathcal{D}$, in order to cover a large set of possible geometric alternatives. The restrictions in $\mathcal{D}$ originate from kinematic and dynamic constraints such as collisions, minimum clearance, joint limits and torque limitations, where the current geometric state is considered for evaluation, including all objects that are grasped by the robot.

An A* planner (Hart et al. 1968) is used to navigate to the closest possible collision-free target position as described in Algorithm 5.3. Depending on the object the appropriate ROI $\boldsymbol{\varphi}$ is selected. The A* planner searches for the shortest path $\boldsymbol{X}$ to reach this goal in the map $\boldsymbol{\lambda}$, which is represented by a discretized grid. The robot cannot reach the exact goal position as it is derived from the origin of the target

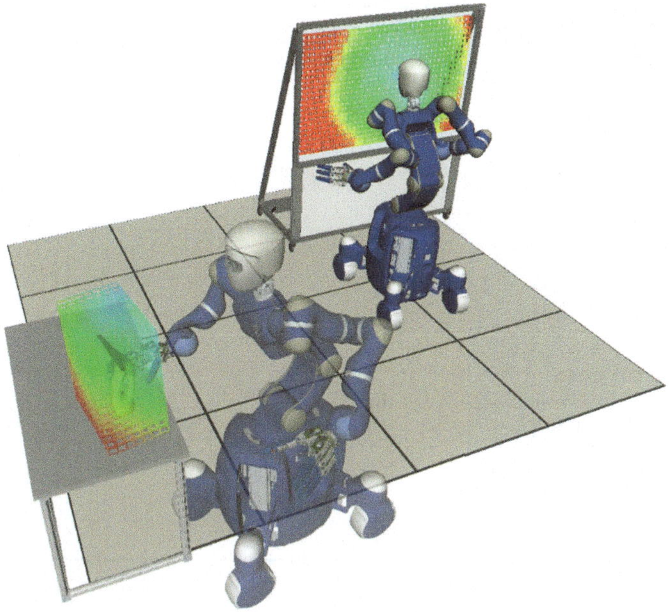

Fig. 5.7 Optimal intersection of the capability map for the right manipulator and the ROI of a window wiper on the left, and the optimal intersection for the ROI of a window on the right

object and therefore fully occupied. The map is thus initially empty and the planning algorithm is used to explore the map. Therefore, the path has to be checked for collisions afterwards. In case of collisions, the respective map positions are marked as obstacles. If the robot is close enough to reach the object ($r_{roi} \geq 0.1$), the final position $\boldsymbol{q}_{end}$ is optimized as shown in (5.2). The procedure is repeated until a feasible path is found or the complete map is explored without success.

Reasoning about navigation is integrated into the geometric planning loop listed in Algorithm 5.2. It does not appear in the symbolic transition. Although it would be possible to add an additional `(reachable ?o ?m)` precondition to the respective Action Templates this would generate unnecessary overhead as it is illustrated in 5.1.

Listing 5.1 PDDL definition for a reachable pick action

```
(:action _object.pick:
    :parameters (?o - _object ?m - _manipulator
         ?s - _surface)
    :precondition (and (free ?m) (on ?o ?s)
         (reachable ?o ?m)))
    :effect (and (bound ?o ?m) (not (free ?m))
         (not (on ?o ?s)))
)
```

Algorithm 5.3: Integrated navigation algorithm.

Input: The object $\mathcal{O}$, providing the Region of Interest φ.
Output: The workspace trajectory $\boldsymbol{X}$.

reached $\leftarrow$ False
$\boldsymbol{\lambda} \leftarrow$ CreateEmptyMap()
$\varphi \leftarrow$ GetROI($\mathcal{O}$)
$\boldsymbol{X} \leftarrow$ List()
while *reached* $\neq$ *True* **do**
    $\boldsymbol{X} \leftarrow$ A*($\boldsymbol{\lambda}$, φ)
    if *CheckCollision*($\boldsymbol{X}$) = *False* **then**
        reached $\leftarrow$ True
    else
        UpdateMap($\boldsymbol{\lambda}$)
        if *CheckReachability*(φ) ≥ 0.1 **then**
            reached $\leftarrow$ True
$\boldsymbol{x}_{\text{end}} \leftarrow$ OptimizeReachability(φ)
$\boldsymbol{X}^\frown \langle \boldsymbol{x}_{\text{end}} \rangle$
return $\boldsymbol{X}$

This predicate would force a robot to navigate to a certain object in every symbolic step. However, there is no geometric feedback in pure symbolic planning to reason about this decision. Additionally the generality of the symbolic domain decreases since the extra precondition has to be considered for every action. Using hybrid reasoning, this decision can be postponed to the geometric reasoning step. Instead of directly executing an Action Template, the reachability for the action-related object is first validated w. r. t. the object ROI. The navigation function as described in Algorithm 5.3 is executed when the reachability is beneath a minimal reachability threshold of $r_{\text{min}} = 0.5\,r_{\text{max}}$, which means that the object must be at least 50 % reachable by the robotic manipulator. After the object is within reach, the Action Template is parsed and executed. No symbolic overhead is generated this way. A seamless integration of the navigation module within the hybrid reasoning procedure is achieved.

5.2.2 *Whole-Body Motion Planning*

The initial navigation toward an optimal base position is not always sufficient to solve a manipulation tasks. Especially wiping motions often extend over wide areas. In this case it is necessary to move the base of the robot coordinated with the upper body and the arms, in order to reach the entire target surface. For example, with a width of 1.5 m and a height of 1.0 m, the window pane of the window introduced in Fig. 5.7 is too large to follow the Cartesian task motion by only using the arm of the robot. Even with the aid of the torso it is not possible to maintain the contact with

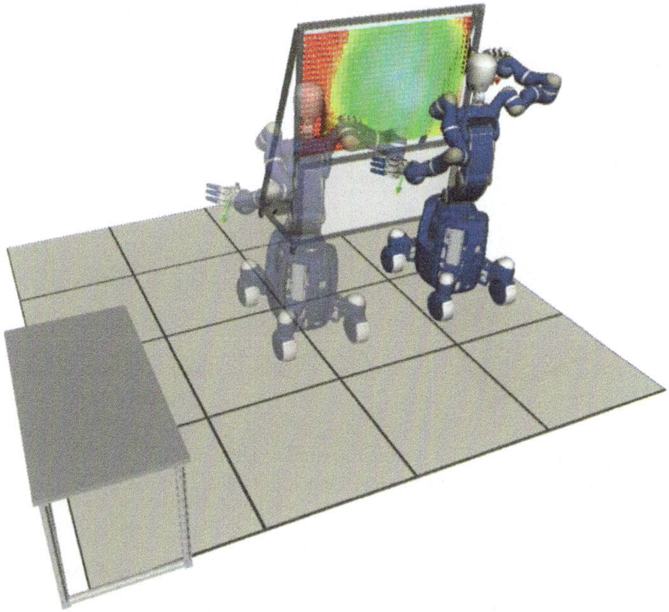

Fig. 5.8 Simulation of whole-body motions to solve the window wiping task. The robot executes wiping motions from the left to the right, starting at the top of the window pane. The large area requires the robot to use its right arm, the torso and the mobile base to reach the entire window

the window along the complete trajectory. Especially in the corners, the reachability index decreases until the task gets unfeasible, not to mention the reduced capability to compensate for possible disturbances during task execution. Consequently, the mobile base is mandatory to accomplish the task. The additional degrees of freedom of the mobile base can be considered by the applied discrete inverse kinematics solver. The additional DOF are thereby added based on the Action Template logic. This way, the base follows the lateral movement of the end-effector, respectively the window wiper, along the window pane to extend the workspace of the robot as it is visualized in Fig. 5.8. As one can see, the robot moves to the left to shift the reachable workspace of the right manipulator. Otherwise, the left hand side of the window would be outside the initial reachability (red area).

Two plots in Fig. 5.9 illustrate the task execution. The first plot (left) shows the commanded six-dimensional task trajectory of the window wiper blade. It correlates to the motions visualized in Fig. 5.8. It is represented as overlay on the simulated window pane. Blue segments are touching the window. Green segments have no contact. The second plot (right) illustrates the reachability index r for the end-effector position of the right manipulator for all wiping motions along the window pane. The solid, black plot indicates the discrete reachability index r. It is close to the maximum value r_{max} (blue, chain-dotted) and never below $r_{\text{min}} = 0.5\, r_{\text{max}}$ (green, dashed) along the trajectory due to optimal base position and torso configuration. For comparison,

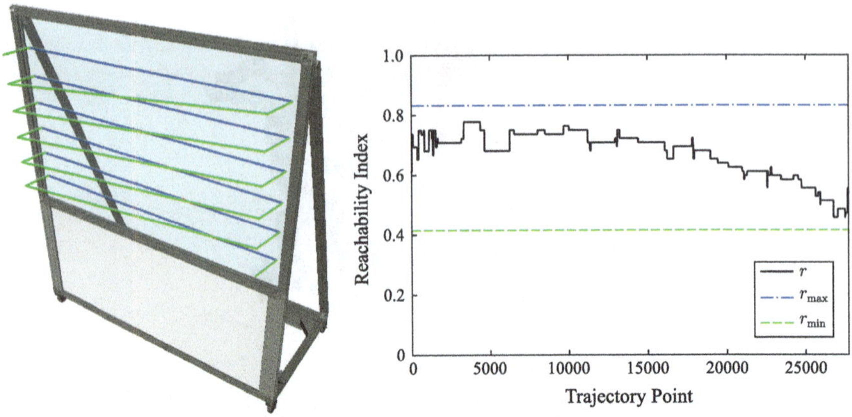

Fig. 5.9 The workspace trajectory for wiping the window (left) and the corresponding reachability index r for the end-effector along the complete trajectory (right)

the colors for the thresholds correspond to the color scheme used for visualizing the Capability Map in Fig. 5.6 to Fig. 5.8.

It can be observed that the reachability is relatively high along the complete trajectory. Maintaining such a high reachability provides more space for reactivity, potential evasion movements, disturbances to be compensated, and additional control tasks that have to be executed. This is achieved due to the optimal positioning of the robot w. r. t. the maximized reachability r_{roi} before the actual execution. The reachability index decreases over time since the lower part of the window pane is harder to reach for the robotic manipulator. Even though the task maintains theoretically feasible unless the reachability index converges to zero, it is notable that this is only achieved due to the fact that the lateral motion along the window pane is supported by the mobile base of the robot.

5.2.3 *Discussion*

Capability Maps provide a powerful tool to bridge the gap between symbolic planning and mobile manipulation as it was intended by Zacharias et al. (2007). The representation can be accessed quickly and is therefore suitable to query larger goal regions in order to abstract away from distinct goal positions toward whole ROIs. Optimizing the reachability for an initial manipulation position is thereby often sufficient to reach a specific goal region from different orientations and with additional space to respond to unforeseen events such as external disturbance. However, the investigations in this section are only of preventive nature. Maintaining a high reachability along a trajectory indicates a high reactive potential, yet it is not guaranteed that joint limits and singularities can be circumvented while generating joint motions

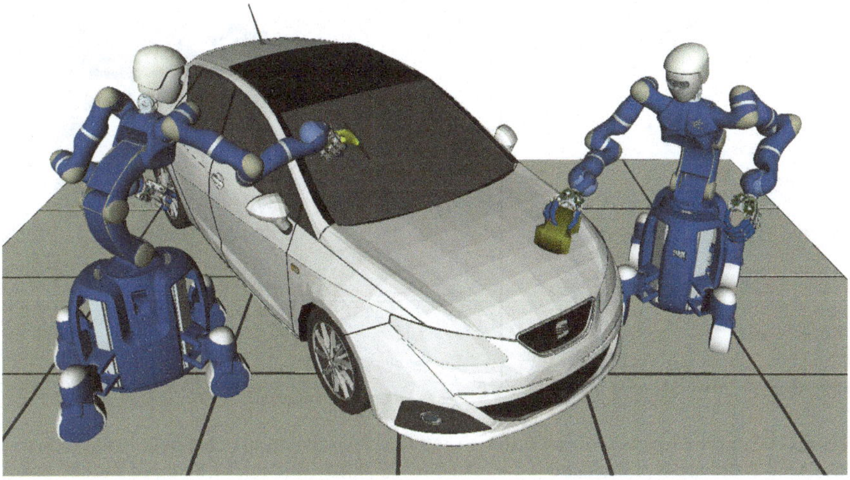

Fig. 5.10 Illustration of the robotic agent Rollin' Justin skimming the windshield of a car and scrubbing the hood with a sponge

w. r. t. a predefined workspace trajectory. Using a discrete inverse kinematics to calculate global joint motions may thus still lead to local minima since both planning and execution are based on local methods. Moreover, compliant execution of the motions may lead to different joint states, possibly resulting in singularities or joint limits. Huaman and Stilman (2012) resolve joint motions with a given Cartesian trajectory based on a discretized Jacobian null space and a backtracking mechanism to prevent local minima and additionally avoid obstacles. A similar approach is therefore investigated in the following section to extend the workspace of robotic manipulators beyond the boundaries of Capability Maps.

5.3 Effect-Space Planning of Wiping Motions

The research roadmaps for autonomous robots (Christensen et al. 2013; Liepert et al. 2014) envisage that robotic agents manufacture products in close physical cooperation with their human co-workers or in the more distant future as helpers that enable old or disabled citizens to live their lives more independently. Such robotic agents would have to be capable of taking over a substantial part of the human daily household chores. As already emphasized, wiping tasks are thereby most frequently observed. Accordingly, generic methods to cope with arbitrary instances of wiping tasks are mandatory. It is not sufficient to hardcode wiping motions as it was done in Sect. 5.1.1.

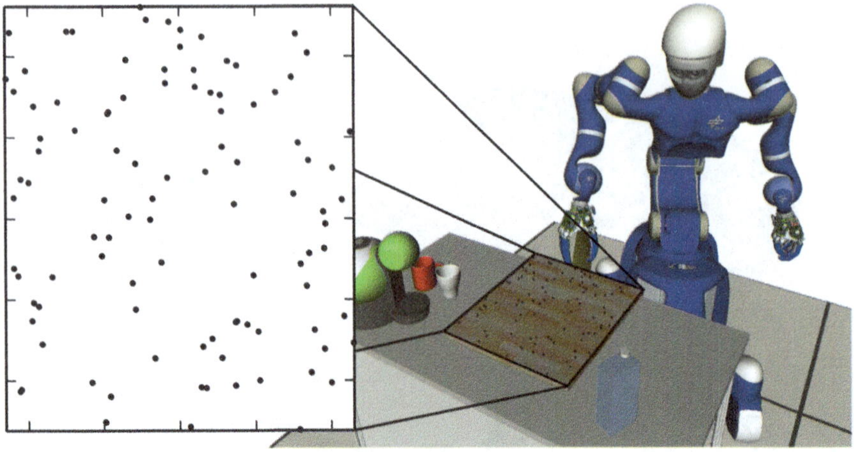

Fig. 5.11 Simulated particle distribution on the target surface in the chopping board scenario. The initial distribution of the particles is randomly sampled. This can be assumed for any barely perceivable medium, such as small breadcrumbs

With reference to the classification conducted in Chap. 3, the nature of wiping tasks is defined w. r. t. the relation between the tool, the surface, and the medium to be manipulated. The tool orientation, and the direction of motion are thereby crucial aspects for the task performance. For example, cleaning the body of a car can be solved by repeatedly moving a sponge in random orientations and directions along the car surface. Skimming snow from a windshield is only successfully if each wiping motion is directed toward the edge of the window, having the tool properly aligned as it is illustrated in Fig. 5.10.

One of the most essential aspect of a manipulation action is its effect to the environment. Automated planning systems take this into account by describing actions based on their pre-conditions and effects. Similarly, it is desirable for a robot to plan actions w. r. t. the anticipated effect, i.e. directly in the *effect-space*. The effects of wiping actions cannot be described based on discrete changes to a single object. As described in Chap. 3, it is the effect to an abstractly defined medium that is the essence of these actions. Humans possess a general understanding for almost any task. This makes it possible to imagine a procedure they have never seen before based on verbal communication and common sense reasoning (Hayes 1978). This abstraction from symbolic descriptions toward the effect-space of an action is probably the most powerful cognitive ability of humans, unparalleled in its nature. Accordingly, effect-space planning is most desirable for the evolution of cognition-enabled service robots as it is discussed in the following sections.

The effect-space for wiping motions is represented by the particle distribution model introduced in Sect. 4.3. The first step (Sect. 5.3.1) to plan wiping motions in this space is to cover the target area, i.e. distribute a set of graph nodes n_i and developed

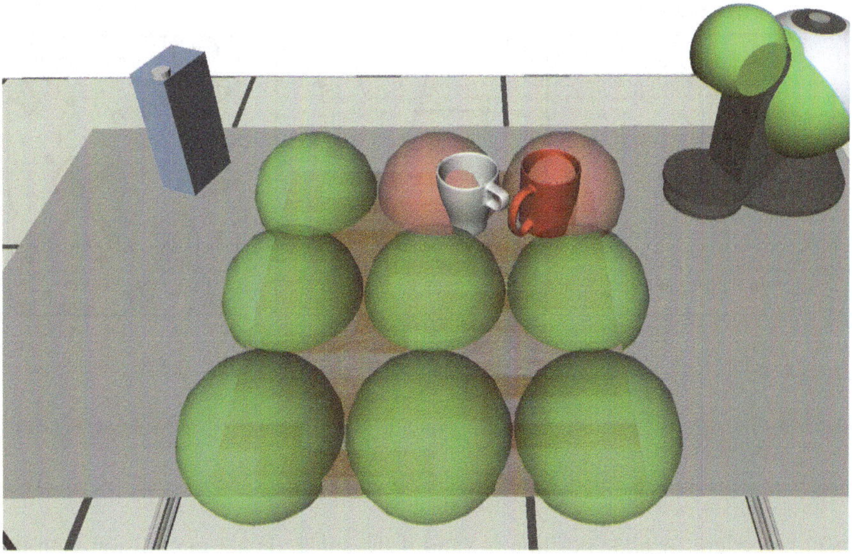

Fig. 5.12 The chopping board defines the target surface and thus the boundaries for the particle distribution. The visualized collision sphere matches the size of the sponge in Fig. 4.6. It is visualized for the graph node positions n_i, where green is valid and red in collision with the environment

the graph according to the semantic goal state. In a second step (Sect. 5.3.2), the framework generates the matching whole-body robot motions.

5.3.1 Reasoning About Cartesian Wiping Motions

The reasoning algorithms developed in this section are based on the particle distribution model $\mathcal{P}$ and the so called *Semantic Directed Graphs* $\mathbf{SDG} = f(\mathcal{P}, \mathcal{S}_g, \mathcal{G}_{t_0})$, which are represent wiping motions based on the symbolic goal state $\mathcal{S}_g$ and the volumetric model of the environment $\mathcal{G}_{t_0}$. As introduced in Sect. 4.4, SDGs are based on the fact that wiping motions can be considered as coverage path planning problem. An example for this problem is shown in Fig. 5.11, where the chopping board provides the target surface.

The node distribution for the surface coverage is restricted by the current state of the geometric environment $\mathcal{G}_{t_0}$, i.e. the volumetric model for geometric planning of wiping motions. Each node of the graph covers a region that is equivalent to the dimensions of the tool. The dimension is approximated by a *spherical model*, which is implemented to explore the target surface as shown in Fig. 5.12. The spherical model is utilized to validate hypothetical node positions during the graph develop-

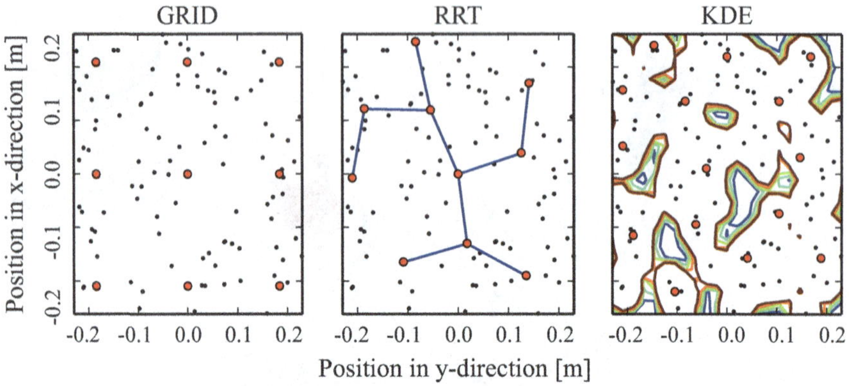

Fig. 5.13 The coverage strategies utilized to explore the target surface. The visualized area corresponds to the obstacle free chopping board surface

ment procedure. A node position is valid if the sphere is not in collision with the environment (except for collisions with the robot, the tool, and the target surface). The collision sphere diameter is defined as

$$d_s = ||\boldsymbol{T}_{aoe}||, \tag{5.3}$$

where $\boldsymbol{T}_{aoe}$ is the area of effect of the applied tool only (e. g. the nozzle of a vacuum cleaner). The full extent of the tool is not directly considered, as this might be much bigger than the actual required space. For example, moving the nozzle of a vacuum cleaner under a bed is sufficient to absorb the dust. The task can be accomplished although the vacuum cleaner itself is too big to fit under the bed.

Three different coverage strategies have been investigated, namely a discretized grid (GRID), Rapidly Exploring Random Trees (RRT) (LaValle 1998), and a Kernel Density Estimation (KDE). The methods are compared in Fig. 5.13, where red dots mark the resulting graph nodes.

Discretized Grid

The first coverage strategy constitutes a simple grid heuristic. The radius, $d_s/2$, of the collision sphere is used to calculate the grid resolution within the bounds of the target area. The grid-based strategy is uninformed and may be applied if no prior knowledge on the particle distribution is available.

Rapidly Exploring Random Trees

RRT (LaValle 1998) is a well established method in research on path planning and exploration. The algorithm samples a random configuration $\boldsymbol{q}_{rand}$ in the free space C, calculates the nearest neighbor $\boldsymbol{q}_{near}$, and extends the tree starting from this configuration toward $\boldsymbol{q}_{new}$, which incorporates the maximal expansion length $\boldsymbol{q}_{delta}$. For the approach, $\boldsymbol{q} \in \mathbb{R}^2$ and $\boldsymbol{q}_{delta} = d_s/2$. The algorithm is biased to explore

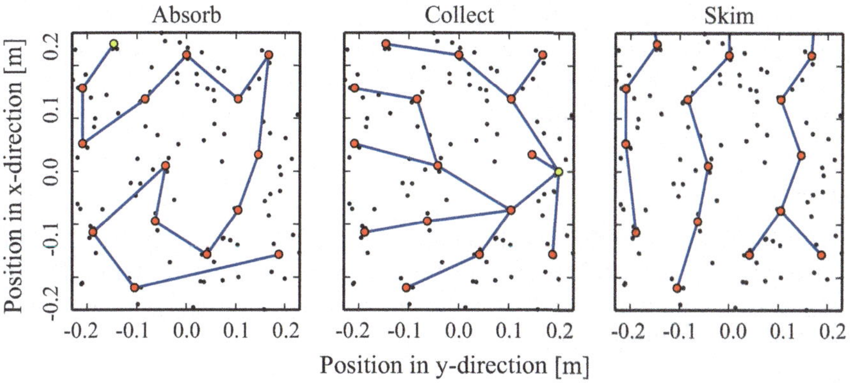

Fig. 5.14 Top view of the developed graph structures for the three prototypical removal actions absorb, collect, and skim. The goal nodes are marked in yellow. The goal nodes for the skim action are located at the top, outside the boundaries of the plot

uncovered regions and it is therefore predisposed for the region coverage problem. The RRT can be augmented to reject nodes that are too far away from the particles to include prior knowledge about the particle distribution.

Kernel Density Estimation

The third strategy to be investigated is a Gaussian KDE to estimate particle probability regions within the particle distribution $\mathcal{P}$:

$$\boldsymbol{K}(x) = \frac{1}{\mathrm{N}} \sum_{i=1}^{\mathrm{N}} e^{-||\mathcal{P}_i - x||^2/h^2}, \tag{5.4}$$

where h is the bandwidth of the kernels. The multivariate KDE is visualized as a contour plot on the right in Fig. 5.13. This continuous representation is used to select the most significant peaks with a clearance of d_s, which places the nodes naturally at the position with the highest effect. This approach is most beneficial if prior knowledge about the distribution is available, e. g. perceived by a vision system, or estimated based on haptic feedback (see Chap. 7).

The calculated nodes n_i serve as the starting point to grow SDGs with different semantic goals $\boldsymbol{\mathcal{S}}_g$. As the coverage strategy may affect the task performance, it is evaluated how the execution time and the task performance are influenced (based on the goal definitions in Sect. 4.3) in the remainder of this section. In the following the distribution generated by the KDE is used to outline the graph generation for the three removal actions identified in Chap. 3 (compare Fig. 5.14 with Fig. 3.4).

The three removal actions have to be implemented in correspondence with the constraint definitions in Sect. 4.3. In fact, the geometric process models of these actions need to produce the desired effect to the particle distribution $\mathcal{P}$. For example, the desired effect of the absorb action is to remove all particles from the surface, by simply getting in contact with each particle to trigger a delete operation. However, collecting and skimming require directed tool motions to have the particles pushed toward a certain goal area (collect), or pushed from the edge of the surface (skim), respectively. Suitable standard solutions for these issues can be found in the literature on graph theory.

Absorbing

A possible solution for the absorb action is the *Traveling Sales Person (TSP)* algorithm. Hess et al. (2012) showed that this is an effective approach to solve unconstrained robotic wiping tasks in a generalized manner. The outcome is a natural curved motion covering all nodes of the graph.

Collecting

The collect action requires to direct the graph to a certain goal region, i.e. a single root node. In graph theory this problem is described by a *Minimum Spanning Tree (MST)* (Graham and Hell 1985). The distance between the nodes serves thereby as cost function with a maximum connection length $l_{max} = d_s$.

Skimming

Skimming is related to collecting and therefore implemented as multiple MSTs. The goals are represented by multiple goal nodes outside the edges of the target surface. Depending on the object that provides the target surface, only a subset of edges may be valid. For the chopping board, only the edge facing the robot is defined as goal region (right in Fig. 5.14). As a result, multiple trees expand toward the closest goal node.

5.3.2 Generating Whole-Body Joint Motions

The development of SDGs in Cartesian space is only a first estimate for the feasibility of the planned wiping motions. The collision sphere model is utilized to check for collision free translational motion of the TCP along the developed graph structure. However, the orientation of the tool is not considered until now. Similarly, the joint state of the robot has to be integrated into the reasoning process to verify the overall feasibility. It is most desirable to move the tool perpendicular to the planned motion, such that as many particles as possible are effected. However, for some cases it is required to rotate the tool to circumvent collisions or local minima in the joint space. Additionally, the highly redundant robot may sometimes be able to resolve local minima on its own by selecting a different joint configuration beforehand.

The graph structure of an SDG represents Cartesian tool motions that serve as the basis for whole-body robot motions. The underlying problem to resolve a Cartesian path into joint motions is formulated as *path following problem*. For each Cartesian pose $\boldsymbol{x}$ on a Cartesian path $\boldsymbol{X}$, the robot has to find a joint velocity $\dot{\boldsymbol{q}}$

$$\dot{q} = J^{\dagger}\dot{x} + (I - J^{\dagger}J)\dot{q}_0, \tag{5.5}$$

here $J^{\dagger}$ is the generalized inverse of the Jacobian matrix (e. g. the Moore-Penrose pseudo-inverse (Buss 2004)). The left term on the right hand side of the equation minimizes the Euclidean norm of the joint velocity. The right term on the right hand side of the equation exploits the redundancy of the manipulator to satisfy secondary criteria, e. g. avoidance of collisions or local minima. The joint velocity $\dot{q}$ and the joint acceleration $\ddot{q}$ must not exceed the limits of the robotic manipulators. Moreover, the resulting joint path Q must not collide with any obstacle nor the robot itself. These issues are particularly challenging for humanoid robots with a high number of DOF, as the search space grows exponentially with the number of joints.

An inverse kinematics based approach to the path following problem is proposed by Konietschke and Hirzinger (2009). This algorithm is based on non-linear optimizations, which allows Rollin' Justin to track arbitrary Cartesian trajectories on-line. However, the algorithm does not consider collisions with obstacles nor the robot itself. The local method is prone to fall into local minima, such as singularities in the joint space. A global path following approach is proposed by Huaman and Stilman (2012). The authors propose a deterministic approach that exploits the redundancy in the discretized Jacobian null-space of the robotic manipulators. The algorithm implements a breadth-first backwards search procedure. Local minima or obstacles can thereby be circumvented by backtracking to previous joint configurations. The major drawback of this method is high computation time. It is advised to exploit local path following methods whenever possible and utilize global search-based methods only if all local alternatives are revealed to be unfeasible. This poses a trade-off between planning time and completeness while it mitigates the drawbacks of the methods.

Typically, path following algorithms aim to follow a given Cartesian path as exactly as possible, where all six dimensions of the task are tracked (i.e. three translational dimensions and three rotational dimensions). However, this poses unnecessary restrictions for wiping tasks. For instance, a sponge may be rotated along the normal of the target surface to yield better reachability without significantly decreasing the wiping effect. Another example is the cleaning motion of a window wiper which has to be moved orthogonal to the wiper blade in order to achieve the desired effect. However, rotating the wiper along the main axis of the blade does not impair the cleaning result. To this end, a path following method is proposed, which is aware of the free tool DOF available in the Cartesian space.

The overall path following procedure to plan wiping motions w. r. t. the redundancy of the robot and the Cartesian space is defined as follows. Initially, the local path following method described by Konietschke and Hirzinger (2009) is applied, with additional checks for collision between the robot and the environment to follow the Cartesian tool motions of the SDG. The sequence in which the SDG branches are processed depends on the wiping action. The absorb motions consists of a single path and is therefore unambiguous. The branches developed for the collect and skim actions are resolved iteratively, starting from the leaf node closest to the initial tool pose. The motion propagates backwards to the root node of the respective branch.

Algorithm 5.4: The path following algorithm to resolve SDGs in joint space.

Input: The initial joint configuration $\boldsymbol{q}_{n_i}$, the goal node n_{i+1}, and the step-size δ
Output: A continous joint path $\boldsymbol{Q}$

$\boldsymbol{x}_{\text{start}} \leftarrow$ CalculateToolPose($\boldsymbol{q}_{n_i}$, $\boldsymbol{x}_{\text{grasp}}$)
foreach $\boldsymbol{x}_{goal} \in$ *IterateFreeDOF(n_{i+1})* **do**
  $\boldsymbol{Q} \leftarrow$ List()
  $\boldsymbol{X} \leftarrow$ Interpolate($\boldsymbol{x}_{\text{start}}$, $\boldsymbol{x}_{\text{goal}}$, δ)
  foreach $\boldsymbol{x}_i \in \boldsymbol{X}$ **do**
    $\boldsymbol{x}_{\text{eef},i} \leftarrow \boldsymbol{x}_i \cdot \boldsymbol{x}_{\text{grasp}}$
    $\boldsymbol{q}_i \leftarrow$ FindIK($\boldsymbol{x}_{\text{eef},i}$)
    if *IsValid($\boldsymbol{q}_i$)* **then**
      $\boldsymbol{Q} ^\frown \langle \boldsymbol{q}_i \rangle$
    else
      break
  if *Length($\boldsymbol{Q}$) = Length($\boldsymbol{X}$)* **then**
    return $\boldsymbol{Q}$
return List()

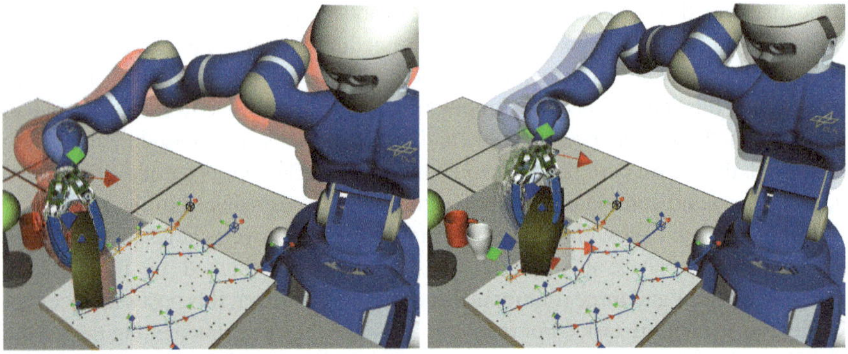

Fig. 5.15 Visualization of the tool orientation in case of collision and no reachability. Initially, each node n_i is oriented toward the next node n_{i+1}. The tool rotation is interpolated along the Cartesian path, i.e. the edge (n_i, n_{i+1}) between this nodes. If a configuration along this path is in collision (left), or unreachable (right), the free tool DOF are exploited

The procedure to generate continuous joint motions between two nodes is outlined in Algorithm 5.4. The TCP poses are derived from the nodes in the graph. Each node n_i is oriented toward the next node n_{i+1} in the branch. The resulting poses (i.e. the translation and orientation) for the nodes are utilized to calculate the initial hypothesis for the start pose $\boldsymbol{x}_{\text{start}}$ and the goal pose $\boldsymbol{x}_{\text{goal}}$ of the TCP as formulated in (5.5). The edge (n_i, n_{i+1}) in between the nodes is interpolated to resolve the constraints of the path following task. Ideally, all interpolated poses $\boldsymbol{x}_i$ are reachable and collision free so that the robot can manipulate the tool accordingly[2]. However, in cluttered environments it is likely that some of the edges in between two nodes may be blocked by obstacles. Additionally, the robot may face local minima (see Fig. 5.15). In either case it is backtracked to the initial pose $\boldsymbol{x}_{\text{start}}$, where an an alternative goal pose $\boldsymbol{x}_{\text{goal}}$ is selected w. r. t. the free DOF of the tool. The procedure is repeated until all alternative combinations in the discretized search space of the free DOF are evaluated. If no successful path can be found, the global method described by Huaman and Stilman (2012) is applied, which is guaranteed to find a path if one exists (given enough time). If no feasible inverse kinematics solution is found at all, the node is abandoned and the algorithm moves on with the remaining nodes of the branch. It is notable that a Cartesian impedance controller alone would not be suficient to solve the described path following problem, as the controller is not aware of the surrounding obstacles it could collide with.

5.3.3 Reachability Extension

For now, the wiping tasks investigated in this section have been considered as globally realizable in terms of reachability. However, it is quite common for wiping tasks in everyday environments to extend over large areas as emphasized in Sect. 5.2.2. Some example tasks are vacuuming the floor, wiping larger windows, and cleaning whole table surfaces. These tasks require to reposition a mobile manipulator in order to cover the entire task region. This work, proposes to incorporate this issue by introducing *extended Semantic Directed Graphs (eSDG)*, which augment the graph nodes n_i with reachability information for the robotic manipulator. This information is used to reposition the base of the robot during the manipulation planing procedure. The reachability analysis introduced in Sect. 5.2.1 is adapted in this section by augmenting the reachability information for each graph node in the eSDG structure. This allows the robot to optimize its base positions for smaller sub-graphs of the eSDG and the underlying particle distribution model, respectively.

The sub-graphs of eSDGs, reffered to here as cluster, are computed according to Algorithm 5.5. Initially, there is one cluster $\boldsymbol{C}$ for all nodes. The robot base position $\boldsymbol{x}_{\text{base}}$ is optimized w. r. t. the mean reachability for the nodes in the cluster (5.2). For each node n_i, the reachability indices r_i are calculated under consideration of the free DOF provided by the tool (see Algorithm 5.4). Based on the earlier investigations (see Sect. 5.2.1), an area is considered as sufficiently reachable if the mean reachability for the involved poses is higher than $0.5\, r_{\max}$. The cluster and the corresponding base position are stored if all nodes are sufficiently reachable. If a sub-set of nodes is unreachable, the K-Means algorithm is applied to split the cluster into two new clusters ($K = 2$). The two cluster centers are initialized with the center of the reachable nodes $\boldsymbol{R}$, and unreachable nodes $\boldsymbol{U}$, respectively. An alternative approach would

---

[2]The default orientation $\boldsymbol{x}_{\text{goal},0}$ is most effective w. r. t. the manipulation of the medium, i.e. the particles.

Algorithm 5.5: Extended SDG function, ExtendSDG($\boldsymbol{C}$, **eSDG**).

Input: The initial cluster $\boldsymbol{C}$, and the **eSDG** structure
Output: The **eSDG** structure, including a list of clusters $\boldsymbol{C}$ and their base positions $\boldsymbol{x}_{base}$

$\boldsymbol{x}_{\text{base}} \leftarrow$ OptimizeBasePose($\boldsymbol{C}$)
$\boldsymbol{R} \leftarrow List()$
$\boldsymbol{U} \leftarrow List()$
foreach $n_i \in \boldsymbol{C}$ **do**
    $r_i \leftarrow$ CalculateReachability(n_i, $\boldsymbol{x}_{\text{grasp}}$)
    if $r_i >= 0.5\, r_{max}$ **then**
        $\boldsymbol{R}[i] \leftarrow n_i$
    else
        $\boldsymbol{U}[i] \leftarrow n_i$
if $\boldsymbol{U} = \emptyset$ **then**
    eSDG $\frown \langle \boldsymbol{C}, \boldsymbol{x}_{\text{base}} \rangle$
else
    foreach $\boldsymbol{C}_i \in$ *KMeans(**C**, Centers(**R**, **U**))* **do**
        eSDG $\leftarrow$ ExtendSDG($\boldsymbol{C}_i$, **eSDG**)
return eSDG

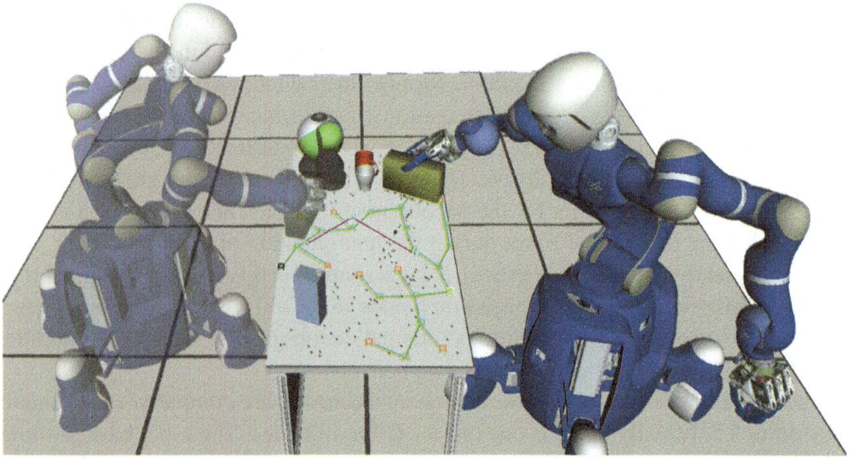

Fig. 5.16 The eSDG collect graph consists of three clusters visualized by the cyan nodes, connected by purple lines. As the goal is located on the right (yellow node), the cluster on the left is resolved first (transparent robot). The A* algorithm is used to navigate between the clusters

be to select the clusters according to the initial reachability estimation, i.e. using $\boldsymbol{R}$ and $\boldsymbol{U}$ directly as new cluster. However, in comparison, the proposed approach yields larger connected areas rather than smaller scattered regions. The algorithm is executed recursively.

The eSDG clusters extend the reasoning procedure for wiping actions, such that each cluster implements the strategy of the main graph, i.e. absorbing, collecting and skimming. Moreover, the individual clusters depend on each other. For the collect action, each cluster implements the collect strategy with an intermediate goal. The clusters have to be executed in the right order, such that all motions propagate toward the goal node of the overall action as it is visualized in Fig. 5.16. The extended procedure for the skim action is more complex. The clusters that are not connected with the goal region (i.e. the edge of the target surface) have to be executed at first. The inner clusters for the skim eSDG can therefore be considered as collect actions. Afterwards the outer clusters are executed, such that all particles are removed from the target surface subsequently. In contrast, the clusters of the absorb action may be executed independently. The A* algorithm (Hart et al. 1968) is applied to navigate in between the base positions of the clusters.

5.3.4 Combined Wiping Actions

In some cases wiping actions may be combined to serve a higher level goal. A common example is the disposal of breadcrumbs into a trash can as it is visualized in Fig. 5.17. If the robot would be commanded to skim all particles into the trash can with a single skim action, the result would not be satisfactorily as the goal region defined by the target surface is not specific enough. In case of the chopping board, the goal region is defined as the edge of the board pointing toward the table edge. Since the trash can is smaller than this area, most of the particles would end up outside the trash can instead of inside as intended.

A possible solution to this is to plan multiple wiping actions. The combination of multiple actions to solve a greater goal was already outlined in Fig. 3.3. In the example at hand, the robot is commanded to execute a collect action in the first place to accumulate all particles next to the trash can. The accumulated particle distribution is used as input for the subsequent skim action by utilizing the KDE method to generate a new SDG. As a result, all particles are skimmed off the chopping board right above the trash can.

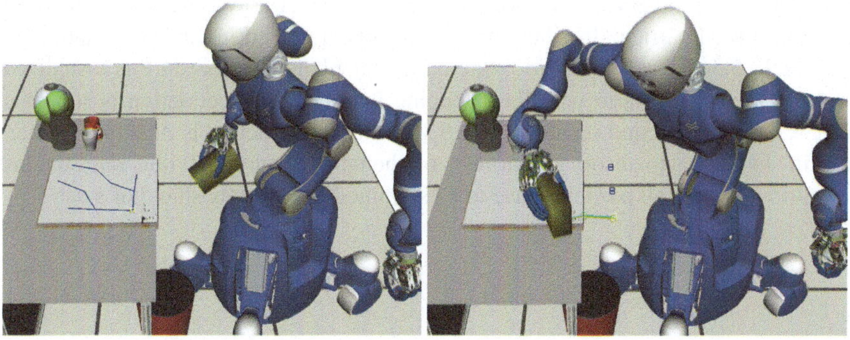

Fig. 5.17 A combination of a collect action (left) and a skim action (right) is used to dispense all particles into a trash can (red)

5.3.5 *Effect Prediction and Evaluation*

The particle distribution model introduced in Sect. 4.3 is utilized to predict the effect of the planned wiping motions in simulation. The performance is rated according to the constraint definitions stated in Sect. 4.3, which are formulated to match the underlying process model of the respective prototypical removal actions encoded as graph structure of the SDG. As the task performance may depend on the initial node distribution on the target surface, this sub-section evaluates the different node distribution strategies w. r. t. the investigated wiping actions. The evaluation is conducted in three different scenarios. Scenario I constitutes the chopping board scenario illustrated in Fig. 4.6. For scenario II, the same environment is assumed without the chopping board, such that the particles are distributed on the entire table surface as shown in Fig. 5.16. Scenario III is a car cleaning scenario, where the target surface is approximated by a planar target surface aligned with the windshield of the car that was introduced with Fig. 5.10.

The sponge introduced with the example wiping task shown in Fig. 5.10 is used in all experiments in order to obtain comparable results w. r. t. the task performance. All actions are considered as unreachable, such that the eSDG strategy applies and the optimization time is incorporated in the evaluation. All three coverage strategies (i.e. GRID, RRT, KDE) are paired with the three removal actions (i.e. absorbing, collecting, skimming) and evaluated w. r. t. the traveled Cartesian distance in contact with the surface, the computation time, the execution time, and the task performance. The calculation of the computation time involves the node distribution, the cluster optimization, and the path following algorithm. The collision check time using OpenRAVE (Diankov 2010) is included for all of these steps. The execution time is based on a maximum velocity of 1 rad/s joint speed, and a maximum joint acceleration of 2 rad/s^2. The maximum base velocity is limited to 0.5 m/s, and a maximum acceleration of 1 m/s^2. The performance measurement is conducted according to the constraint definitions (4.2), (4.3), and (4.4). Accordingly, the metric for the absorb action can be described as the number of deleted particles, the collect action is evaluated based on the number of particles within the radius r_s around the goal node n_{goal}, and the metric for the skim action is the number of particles pushed outside the boundaries of the target surface. All results are based on the average of five trials with different initial particle distributions. The results are listed in Table 5.1, Table 5.2, and Table 5.3. For each scenario, one particular planning procedure is visualized. In the table top scenario with the chopping board, the collect action is shown in Fig. 5.18. In the second scenario (Fig. 5.19), the absorb action is illustrated, where Rollin' Justin has to absorb all particles on the table. Figure 5.20 illustrates the last scenario in which all particles have to be skimmed off the windshield of a car.

Table 5.1 Evaluation of the chopping board scenario

		GRID	RRT	KDE
Absorb	Cart. dist. (m)	1.52	1.41	1.90
	comp. time (s)	45.96	57.12	62.00
	exec. time (s)	18.01	25.92	27.71
	performance (%)	84.25	85.00	95.75
Collect	Cart. dist. (m)	1.54	1.45	1.90
	comp. time (s)	69.89	68.48	72.24
	exec. time (s)	16.17	32.42	43.28
	performance (%)	86.50	86.75	90.50
Skim	Cart. dist. (m)	1.81	1.62	2.01
	comp. time (s)	107.17	44.25	90.57
	exec. time (s)	32.11	26.51	45.81
	performance (%)	88.00	80.25	97.00

5.3.6 Discussion

The proposed particle distribution is a suitable representation to estimate the outcome of wiping motions in compliant contact. It is applicable to predict the effect of wiping motions generated with the different coverage strategies and measure the task performance w. r. t. the constraint definition for the three removal actions. The combination of local and global joint motion generation strategies provides an effective trade-off between planning times and completeness. In combination with the extension toward wide areas, the algorithm is most suitable for solving arbitrary

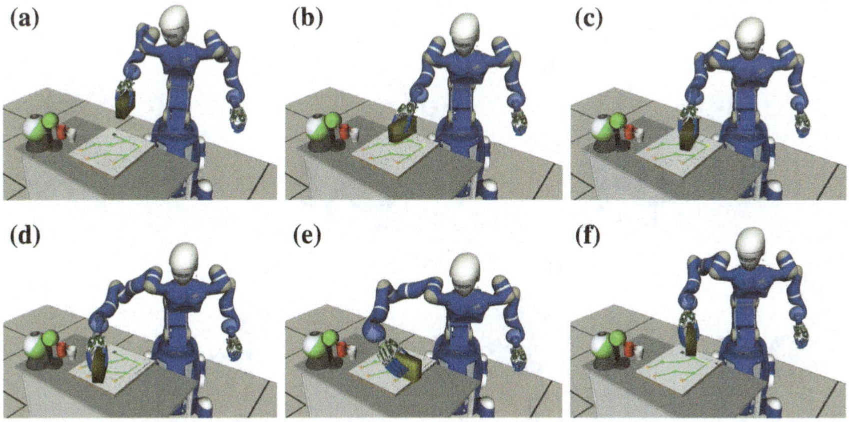

Fig. 5.18 Simulation-based effect prediction for the collect action in the chopping board scenario. The particles contact behavior is visualized over time

Table 5.2 Evaluation of the table top scenario

		GRID	RRT	KDE
Absorb	Cart. dist. (m)	4.34	4.10	3.24
	comp. time (s)	144.47	312.45	169.92
	exec. time (s)	41.30	66.96	52.485
	performance (%)	77.00	79.50	70.50
Collect	Cart. dist. (m)	3.68	3.60	2.92
	comp. time (s)	110.92	161.60	270.56
	exec. time (s)	65.71	83.36	62.14
	performance (%)	70.25	69.25	41.75
Skim	Cart. dist. (m)	5.48	4.95	3.75
	comp. time (s)	211.14	260.34	255.58
	exec. time (s)	113.37	122.76	102.48
	performance (%)	71.75	66.50	53.00

wiping tasks. In conclusion, the approach enables a robotic agent to reason about the task performance of wiping actions as it is aware of the desired state change of the particle distribution. This allows to decide for the most effective strategy to clean a surface given a certain problem instance.

The different coverage strategies vary widely in performance for the individual settings. This allows the robot to reason about the most effective coverage strategy given a concrete problem definition. In case of lesser obstructed environments (as for example posed by the chopping board scenario and the windshield scenario) the KDE mostly outperforms the other methods w. r. t. the performance metrics defined in Sect. 4.3. However, the KDE coverage strategy seems to be biased by obstacles as it is less efficient for the table scenario. In particular, one can observe that the

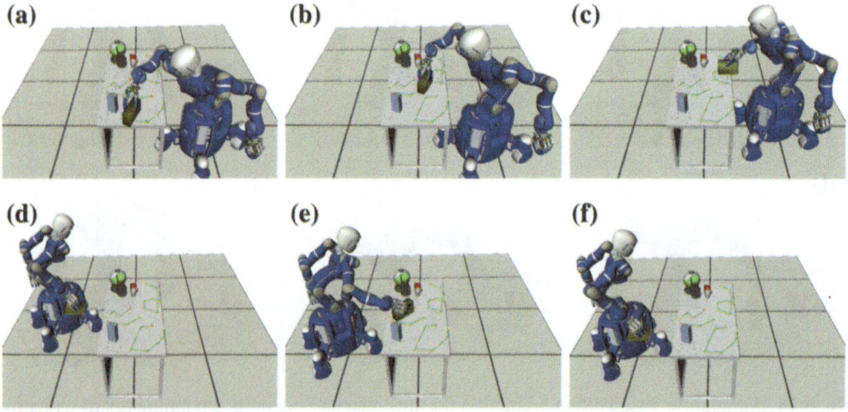

Fig. 5.19 Simulation-based effect prediction for the absorb action in the table scenario. The particles contact behavior is visualized over time

Table 5.3 Evaluation of the windshield scenario

		GRID	RRT	KDE
Absorb	Cart. dist. (m)	4.44	4.84	4.69
	comp. time (s)	314.04	423.21	417.66
	exec. time (s)	77.59	111.90	66.11
	performance (%)	90.13	90.50	91.00
Collect	Cart. dist. (m)	4.97	4.76	4.22
	comp. time (s)	885.45	262.55	428.05
	exec. time (s)	112.47	116.26	97.56
	performance (%)	48.50	39.25	62.75
Skim	Cart. dist. (m)	5.58	5.65	5.89
	comp. time (s)	325.97	212.34	199.03
	exec. time (s)	108.18	103.29	96.03
	performance (%)	94.50	88.50	88.50

graph constructed based on the KDE often results in collisions with the environment which renders the nodes unreachable. This phenomenon is reflected by the poor path length and the low performance. The GRID heuristic and the RRT algorithm perform equally good for most of the scenarios. However, the GRID heuristic is not impaired by probabilistic effects and can thus be used to create repeatable results.

In conclusion it can be stated that the task performance for a specific scenario depends highly on the selected coverage strategy. A robot may have to test all available coverage strategies for a given scenario in order to figure out and execute the most effective one. While this is a drawback for one-time tasks, this observation can be exploited to improve recurring tasks, e. g. industrial manufacturing tasks, such

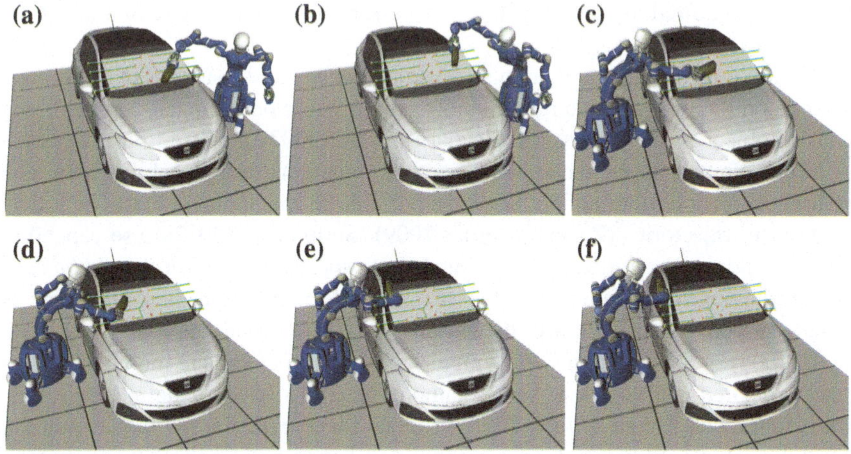

Fig. 5.20 Simulation-based effect prediction for the skim action in the windshield scenario. The particles contact behavior is visualized over time

as polishing the surface of a car. These tasks can be autonomously optimized w. r. t. the execution time or the task performance by iterating over the available coverage strategies and vary the coverage parameters (e. g. r_s). Moreover, the robot can continuously improve on the task performance by integrating episodic memories of previous executions into the reasoning process. For example, the task performance of previous trials can be used to benchmark future trials with alternative task parameters, e. g. a different tool alignment. The generic approach can be utilized to solve previously unseen wiping tasks, only given the desired semantic goal and the geometric properties of the tool and the environment. In this regard, this approach constitutes a valuable addition for autonomous robots.

5.4 Related Work

Hybrid reasoning, by means of a combination of symbolic planning and geometric planning, is mandatory for autonomous solving of manipulation tasks. By only giving a high-level command, a robot has to reason about the symbolic order of the actions to fulfill (e. g. picking a mug before placing it in a cupboard with the available resources), but also about the geometric execution (e. g. how to pick and place in particular with the specific capabilities of the robot).

Several approaches have been proposed to solve manipulation tasks based on combined symbolic and geometric reasoning. Dornhege et al. (2012) ground symbolic actions based on PDDL (Ghallab et al. 1998) by adding semantic attachment modules. Wolfe et al. (2010) directly integrate external geometric solvers into a symbolic Hierarchical Task Network (HTN) planner. Karlsson et al. (2012) invoke geometric actions during the symbolic planning phase and propose geometric backtracking in case of failure. Kaelbling and Lozano-Pérez (2011) focus their work on symbolic and geometric planning under uncertainty for mobile manipulation. However, when mobility is required to solve a task it is necessary to additionally reason about the question where to place a robot.

The topology of a manipulation scenario is crucial for the success of a task. Depending on the position of a robot, objects are possibly out of reach. Manipulating such an object is thus unfeasible. Zacharias et al. (2007) propose to use *Capability Maps* to represent the workspace for robotic manipulators. Capability maps can be used to position the mobile base best possible to manipulate objects or follow workspace trajectories (Zacharias et al. 2009). Stulp et al. (2012a) use capability maps to guide the positioning of a mobile base to so-called *Action-Related Places* to manipulate objects under uncertainty w. r. t. the performed action. *Semantic maps*, as generated from sensor input by Nüchter and Hertzberg (2008) are used by Galindo et al. (2008) in combination with symbolic planning to command mobile robots to reach a desired goal location. Vahrenkamp et al. (2013) propose an *inverse reachability* approach to find suitable base positions for a particular 6D target position. After the robot is positioned most optimally, it has to reason about the geometric process model of the task it is ought to execute.

In the particular case of wiping a surface, this process model is often considered as a *coverage path planning* problem (Latombe 1990), where a robot has to find a path (i.e. for a cleaning device) connecting all nodes of a graph in a time- or effort-optimal way. Okada et al. (2005, 2006) apply an inverse-kinematics-based programming approach to compute whole-body motions for the tasks of sweeping the floor, vacuuming the floor, and washing the dishes with a humanoid robot. Urbanek et al. (2004) demonstrated how machine learning can be used to teach a robot different movement primitives by demonstration in the context of wiping a table. While humans can implement different wiping strategies this way , the meaning behind the motion is not transferred to the robot. Gabriely and Rimon (2001) consider the coverage problem arising for a mobile robot such as a autonomous lawn mower. They propose to sub-divide the search space into a grid and apply variations of the *Spanning Tree Covering (STC)* algorithm to cover the area. Hess et al. (2012) describe an approach to autonomously compute cleaning trajectories for redundant robotic manipulators guiding a sponge on 3d surfaces. They utilize a variation of the *Traveling Salesman Problem (TSP)* and resolve the joint motions of the robotic manipulator by nulls-space optimization in a discretized Jacobian null-space along the Cartesian path (see (Huaman and Stilman 2012)). The approach does not integrate different semantic goals and is therefore only applicable to undirected tasks, such as vacuuming or dusting.

5.5 Summary

The human mind is evolution's most powerful computing machine. It enables humans to ground symbolic task descriptions to meaningful geometric procedures, and provides the means to plan effect-oriented actions for arbitrary problems. In combination, this allows humans to plan everyday manipulation tasks efficiently and effortlessly. This chapter proposes the methods to mirror these human cognitive skills based on machine intelligence. This problem is challenging, but also indispensable to the creation of robots with human-level intelligence. This chapter presented a suitable set of planning methods to solve everyday wiping tasks, which span almost half of all daily household chores Cakmak and Takayama (2013). In particular, it was shown how Action Templates are integrated into an object-centric hybrid reasoning procedure that integrates symbolic planning with geometric planning in a generic way. The concept was extended with mobility reasoning to incorporate tasks that operate over an increased workspace, such as cleaning windows. Moreover, the geometric reasoning procedure defined in the main part of Action Templates was supported by fundamental effect-space planning methods. That is, a concept for goal-oriented planning of wiping motions was proposed based on a particle distribution model. This model was furthermore used to predict the effect of the robot motions qualitatively. In conclusion, the developed methods allow a compliant robot to plan semantically meaningful whole-body wiping motions. The compliant execution strategy shall be detailed in the following chapter.

References

Bejjani, Wissam. 2015. Automated Planning of Whole-body Motions for Everyday Household Chores with a Humanoid Service Robot. Master's thesis, Technical University of Dortmund.

Buss, Samuel R. 2004. Introduction to Inverse Kinematics with Jacobian Transpose, Pseudoinverse and Damped Least Squares Methods. *IEEE Journal of Robotics and Automation* 17 (1–19): 16.

Christensen, Henrik I, et al. 2013. A Roadmap for U.S. Robotics from Internet to Robotics. Technical report, Robotics Virtual Organization.

Cakmak, Maya and Leila Takayama. 2013. Towards a Comprehensive Chore List for Domestic Robots. In *Proceedings of the ACM/IEEE International Conference on Human-robot Interaction (HRI)*, pp. 93–94.

Diankov, Rosen. 2010. *Automated Construction of Robotic Manipulation Programs*. PhD thesis, Carnegie Mellon University, Robotics Institute.

Donald, A. 1980. *Norman and Tim Shallice*. Attention to Action: Willed and Automatic Control of Behavior. Technical report, DTIC Document.

Dornhege, Christian, Patrick Eyerich, Thomas Keller, Sebastian Trüg, Michael Brenner, and Bernhard Nebel. 2012. Semantic Attachments for Domain-independent Planning Systems. In *Towards Service Robots for Everyday Environments*, pp. 99–115. Springer.

Graham, Ronald L., and Pavol Hell. 1985. On the hIstory of The Minimum Spanning Tree Problem. *Annals of the History of Computing* 7 (1): 43–57.

Ghallab, Malik, Adele Howe, Dave Christianson, Drew McDermott, Ashwin Ram, Manuela Veloso, Daniel Weld, and David Wilkins. 1998. PDDL—The Planning Domain Definition Language. *AIPS98 Planning Committee* 78 (4): 1–27.

Galindo, Cipriano, Juan-Antonio Fernández-Madrigal, Javier González, and Alessandro Saffiotti. 2008. Robot Task Planning using Semantic Maps. *Robotics and Autonomous Systems* 56 (11): 955–966.

Gabriely, Yoav, and Elon Rimon. 2001. Spanning-Tree based Coverage of Continuous Areas by a Mobile Robot. *Annals of Mathematics and Artificial Intelligence* 31 (1–4): 77–98.

Harnad, Stevan. 1990. ThE Symbol Grounding Problem. *Physica D: Nonlinear Phenomena* 42 (1–3): 335–346.

Helmert, Malte. 2006. The Fast Downward Planning System. *Journal of Artifcial Intelligence Research* 26: 191–246.

Hart, Peter E., Nils J. Nilsson, and Bertram Raphael. 1968. A Formal Basis for The Heuristic Determination of Minimum Cost Paths. *IEEE Transactions on Systems Science and Cybernetics* 4 (2): 100–107.

Hayes, Patrick J. 1978. *The Naive Physics Manifesto*. Institut pour les études sémantiques et cognitives/Université de Genève.

Hess, Jürgen Michael, Gian Diego Tipaldi, and Wolfram Burgard. 2012. Null Space Optimization for Effective Coverage of 3D Surfaces using Redundant Manipulators. In *Proceedings of the IEEE/RSJ International Conference on Intelligent Robots and Systems (IROS)*, pp. 1923–1928.

Huaman, Ana and Mike Stilman. 2012. Deterministic Motion Planning for Redundant Robots Along End-effector Paths. In *Proceedings of the International Conference on Humanoid Robots (ICHR)*, pp. 785–790.

Konietschke, Rainer and Gerd Hirzinger. 2009. Inverse Kinematics with Closed form Solutions for Highly Redundant Robotic Systems. In *Proceedings of the IEEE International Conference on Robotics and Automation (ICRA)*, pp. 2945–2950. IEEE.

Karlsson, Lars, Julien Bidot, Alessandro Saffiotti, Ulrich Hillenbrand, and Florian Schmidt. 2012. Combining Task and Path Planning for a Humanoid Two-arm Robotic System. In *Proceedings of the International Conference on Automated Planning and Scheduling (ICAPS)*, pp. 13–20.

Kaelbling, Leslie Pack and Tomás Lozano-Pérez. 2011. Hierarchical Task and Motion Planning in the Now. In *Proceedings of the IEEE International Conference on Robotics and Automation (ICRA)*, pp. 1470–1477.

Leidner, Daniel, Christoph Borst, and Gerd Hirzinger. 2012. Things are Made for What They Are: Solving Manipulation Tasks by Using Functional Object Classes. In *Proceedings of the IEEE/RAS International Conference on Humanoid Robots (ICHR)*, pp. 429–435.

Leidner, Daniel, Alexander Dietrich, Michael Beetz, and Alin Albu-Schäffer. 2016b. Knowledge-enabled Parameterization of Whole-body Control Strategies for Compliant Service Robots. *Autonomous Robots (AURO): Special Issue on Whole-Body Control of Contacts and Dynamics for Humanoid Robots* 40 (3): 519–536.

Leidner, Daniel and Christoph Borst. 2013. Hybrid reasoning for Mobile Manipulation based on Object Knowledge. In *Proceedings of the Workshop on AI-Based Robotics at IEEE/RSJ International Conference on Intelligent Robots and Systems (IROS)*.

Leidner, Daniel, Alexander Dietrich, Florian Schmidt, Christoph Borst, and Alin Albu-Schäffer. 2014b. Object-centered Hybrid Reasoning for Whole-body Mobile Manipulation. In *Proceedings of the IEEE International Conference on Robotics and Automation (ICRA)*, pp. 1828–1835.

Leidner, Daniel, Wissam Bejjani, Alin Albu-Schäffer, and Michael Beetz. 2016a. Robotic Agents Representing, Reasoning, and Executing Wiping Tasks for Daily Household Chores. In *Proceedings of the International Conference on Autonomous Agents and Multiagent Systems (AAMAS)*, pp. 1006–1014.

Liepert, Bernd, Stefano Stramigioli, Rainer Bischoff, Uwe Haass, et al. 2014. Multi-annual Roadmap for Robotics in Europe. Technical report, euRobotics Association Internationale Sans But Lucratif (AISBL).

LaValle, Steven M. 1998. Rapidly-Exploring Random Trees: A New Tool for Path Planning.

Latombe, Jean-Claude. 1990. *Robot Motion Planning*. Springer.

Nüchter, Andreas, and Joachim Hertzberg. 2008. Towards Semantic Maps for Mobile Robots. *Robotics and Autonomous Systems* 56 (11): 915–926.

Okada, Kei, Takashi Ogura, Atsushi Haneda, Junya Fujimoto, Fabien Gravot, and Masayuki Inaba. 2005. Humanoid Motion Generation System on HRP2-JSK for Daily Life Environment. In *Proceedings of the IEEE International Conference on Mechatronics and Automation (ICMA)*, pp. 1772–1777.

Okada, Kei, Mitsuharu Kojima, Yuuichi Sagawa, Toshiyuki Ichino, Kenji Sato, and Masayuki Inaba. 2006. Vision Based Behavior Verification System of Humanoid Robot for Daily Environment Tasks. In *Proceedings of the IEEE-RAS International Conference on Humanoid Robots (ICHR)*, pp. 7–12.

Searle, John. 2001. The Chinese Room Argument. *Encyclopedia of Cognitive Science*.

Schulman, John, Alex Lee, Ibrahim Awwal, Henry Bradlow, and Pieter Abbeel. 2013. Finding Locally Optimal, Collision-free Trajectories with Sequential Convex Optimization. In *Proceedings of the Robotics: Science and Systems Conference (RSS)*

Stulp, Freek, Andreas Fedrizzi, Lorenz Mösenlechner, and Michael Beetz. 2012a. Learning and Reasoning with Action-related Places for Robust Mobile Manipulation. *Journal of Artificial Intelligence Research* 43: 1–42.

Urbanek, Holger, Alin Albu-Schäffer, and Patrick van der Smagt. 2004. Learning from Demonstration: Repetitive Movements for Autonomous Service Robotics. In *Proceedings of the IEEE/RSJ International Conference on Intelligent Robots and Systems (IROS)*, pp. 3495–3500.

Vahrenkamp, Nikolaus, Tamim Asfour, and Ruediger Dillmann. 2013. Robot Placement Based on Reachability Inversion. In *Proceedings of the IEEE International Conference on Robotics and Automation (ICRA)*, pp. 1962–1967.

Wolfe, Jason, Bhaskara Marthi, and Stuart J Russell. 2010. Combined Task and Motion Planning for Mobile Manipulation. In *Proceedings of the International Conference on Automated Planning and Scheduling (ICAPS)*, pp. 254–258.

Zacharias, Franziska, Christoph Borst, and Gerd Hirzinger. 2007. Capturing Robot Workspace Structure: Representing Robot Capabilities. In *Proceedings of the IEEE/RSJ International Conference on Intelligent Robots and Systems (IROS)*, pp. 3229–3236.

Zacharias, Franziska, Wolfgang Sepp, Christoph Borst, and Gerd Hirzinger. 2009. Using a Model of the Reachable Workspace to Position Mobile Manipulators for 3-D Trajectories. In *Proceedings of the IEEE/RAS International Conference on Humanoid Robots (ICHR)*, pp. 55–61.

Chapter 6
Compliance Parameterization and Task Execution

This chapter tackles the problem of compliance parameterization and task execution. In particular, it investigates how purely kinematically planned tool motions can be converted into context-aware, physically compliant joint motions of the humanoid robot Rollin' Justin. It describes the concepts and methods that enable a robot to interact compliantly with the environment. It builds on the reasoning methods introduced earlier, providing joint motions for goal-oriented wiping motions. The parameterization and execution of these motions are explained here, in order to achieve the desired effects, i.e. the manipulation of a particular medium. Section 6.1 introduces the fundamental concepts of compliant robot controll in general and impedance control in particular. Section 6.2 builds on these concepts to show how a whole-body impedance controller can be integrated into the reasoning framework developed in Chap. 5. The method is evaluated in Sect. 6.3. Altogether, this chapter describes the third part of the Intelligent Physical Compliance concept, i.e. the execution of compliant manipulation tasks with autonomous robots as it is repeated in Fig. 6.1.

The findings formulated in this chapter are mainly based on three articles. The integration of low-level whole-body impedance control into the high-level AI reasoning framework was outlined in (Leidner et al. 2014b) and later refined in (Leidner et al. 2016b; Ott et al. 2015). The compliant interaction strategies have been mainly developed in (Leidner et al. 2016b).

6.1 Control Strategies for Compliant Interaction

Soft physical interaction is the principal characteristic of many household chores and industrial manufacturing tasks. Humans use their sense of touch and haptic perception to detect contact with the world and the relaxation of their muscles to react

D. S. Leidner, *Cognitive Reasoning for Compliant Robot Manipulation*, Springer Tracts in Advanced Robotics 127,
https://doi.org/10.1007/978-3-030-04858-7_6

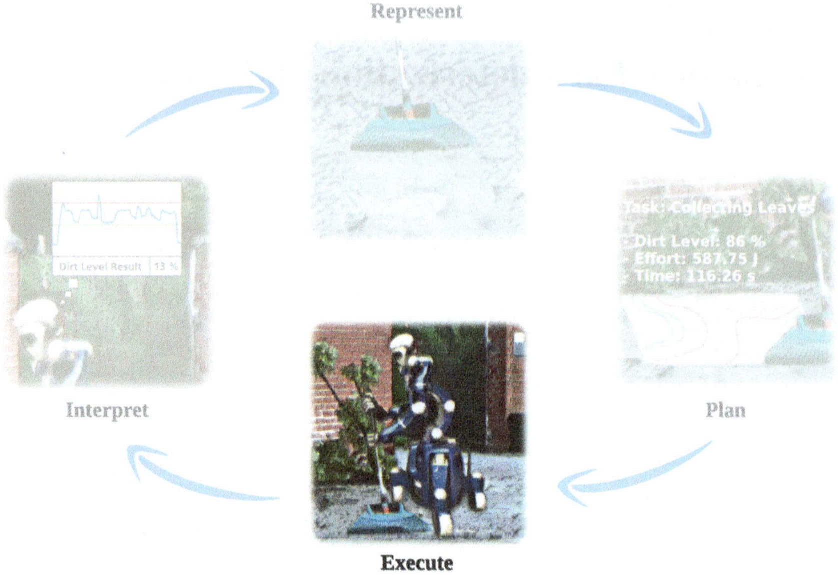

Fig. 6.1 Executing Intelligent Physical Compliance

compliantly. The stiffness of the muscles is adapted w.r.t. the expected or actual contact situation (Grebenstein 2012). For example, when the impact force is not exactly known (e.g. for unforeseen collisions) a human would stiffen all involved muscles in order to prevent tissue damage. In contrast, humans react elastically in case of foreseen impacts (e.g. when landing after a jump). This concept is commonly referred to as *variable stiffness*. In robotics, variable stiffness can be realized through sensory feedback control (active) or with mechanically compliant components (passive). Accordingly, a robot can be capable of *active compliance* or *passive compliance*, respectively.

Active compliance can be effectively realized by means of integrated joint torque sensors. These sensors provide modern light-weight robots with the ability of compliant interaction based on torque control strategies. Especially the concept of *impedance control* (Hogan 1985) poses an interesting concept for active compliance, as it matches the requirements of robotic wiping tasks (Hogan 1987). The impedance control idea is based on the concept of virtual springs. A manipulator will deviate from its motion in the presence of external forces and torques. The extent of this deviation is determined by a user-defined, virtual stiffness. Rather stiff robots, like the LWR III[1] (Hirzinger et al. 2002), can this way interact compliantly with their environment.

Passive compliance is often realized by means of *Variable Stiffness Actuators (VSA)* that contain real mechanical springs and auxiliary drives to modify their stiff-

[1]The LWR III is constructed from harmonic drives. These gears are intrinsically elastic. However, the elasticity arising from this component is of minor magnitude and thus neglected here.

ness. The DLR Hand Arm System (Grebenstein et al. 2011), for example, is a flexible anthropomorphic robot arm realized based on this principle (Friedl et al. 2011; Wolf et al. 2011). In comparison to the DLR LWR III, it is intrinsically elastic as it is designed based on flexible elements. The springs in the actuators allow for compliant manipulation under rough conditions. As an example, the Hand Arm System is able to operate a heavy pneumatic drill to perform a drilling action. The elasticity of the system compensates for the vibrations and impacts such that the actuators remain undamaged.

This chapter focuses on the concept of active compliance, as all experiments are conducted with the humanoid robot Rollin' Justin which is assembled from two LWR III. Most approaches for active compliance control rely on a profound knowledge of the robots kinematics and dynamics. The kinematics for robotic manipulators is typically described by the joint coordinates $\boldsymbol{q} \in \mathbb{R}^{\mathrm{N}}$, where N is the number of joints or actuated DOF, respectively. The workspace of the robot (in control theory, typically referred to as the *operational space* (Khatib 1987)) is spanned by the joint coordinates and can be formulated as $\boldsymbol{x}(\boldsymbol{q}) \in \mathbb{R}^{\mathrm{M}}$, where M refers to the dimensions of the operational space, e.g. 6 for the complete Cartesian space at the end-effector. Humanoid service robots such as Rollin' Justin exhibit a large number of DOF. If $\mathrm{M} < \mathrm{N}$, the robot is kinematically redundant, i.e. it possesses more joints than required to reach a given operational-space task. This redundancy can be utilized to solve additional control tasks in the so-called *null space* of the robot (see Sect. 6.1.2). In general, the relation between the joint configuration $\boldsymbol{q}$ and the operational-space coordinates $\boldsymbol{x}$ of a robot can be expressed by forward kinematics. The inverse kinematics describe the mapping from the operational space $\boldsymbol{x}$ to the joint space $\boldsymbol{q}$. For redundant robots, the ambiguity has to be resolved in order to make this mapping unique, as it was explained in (5.5). While forward kinematics and inverse kinematics describe the relation of joint-space positions and operational-space positions, the relation between joint space torques $\boldsymbol{\tau}$ and generalized, operational-space forces $\boldsymbol{f}$ (including forces and torques) was not explicitly mentioned in the previous chapters. In general, this relation can be formulated as

$$\boldsymbol{\tau} = \boldsymbol{J}(\boldsymbol{q})^T \boldsymbol{f}, \tag{6.1}$$

where $\boldsymbol{J}(\boldsymbol{q})$ is the Jacobian matrix. However, it is addressed by Hogan (1985) that the interaction between two physical systems cannot be formulated based on either the position *or* the force alone. Accordingly, an operational space controller for compliant physical interaction has to incorporate both positions *and* forces in one control law. One possible way to achieve this is presented with the concept of impedance control.

6.1.1 Impedance Control

Impedance control describes an approach to map velocities to forces, by altering the mechanical impedance of a robotic manipulator by means of active control (Hogan 1985). This becomes necessary if the robot has to interact compliantly with its environment. It was interpreted by Hogan that this formulation is a natural consequence of the way physically interacting systems should be represented. In particular, Hogan argues that the environment should be represented as an admittance, which describes a mapping from forces to positions, and the robot should be represented by an impedance, describing the inverse model.

The regulation case for a Cartesian impedance controller is described here based on the illustration of Dietrich (2015), visualized in Fig. 6.2. The input for the operational space controller $\boldsymbol{x}_{\mathrm{des}}$ is here treated as a Cartesian end-effector coordinate, which is introduced as the spatial, virtual equilibrium point in Fig. 1.7. Deviations at the end-effector $\boldsymbol{x} - \boldsymbol{x}_{\mathrm{des}}$ result in forces $\boldsymbol{f}_{\mathrm{cmd}} \in \mathbb{R}^{\mathrm{M}}$ exerted by the robot. In case of physical contact with the environment, the virtual equilibrium point cannot be reached, and a residual external force $\boldsymbol{f}_{\mathrm{ext}}$ remains. The end-effector position $\boldsymbol{x}$ and the end-effector velocity $\dot{\boldsymbol{x}}$ are used as feedback. As a result of the contact situation, the external force counteracts the controller force $\boldsymbol{f}_{\mathrm{cmd}}$. The transposed Jacobian matrix maps this force to a torque command $\boldsymbol{\tau}_{\mathrm{cmd}}$. The torque controller output is denoteted as the motor torque $\boldsymbol{\tau}_{\mathrm{m}}$. The compliant behavior is realized by feedback of $\boldsymbol{\tau}$ in the inner control loop.

In conclusion, one can observe that the inner control loop is the compliant part, while the outer control loop increases the stiffness of the overall closed-loop system. Based on this concept, small- to medium-level stiffness settings can be adjusted. However, further increasing the stiffness toward higher levels results in eventual destabilization. This range of operation makes impedance control the perfect controller for compliant manipulation tasks that require limited positioning accuracy as observed for the example of wiping chores. By additionally clamping the controller force to a certain maximum, it is furthermore possible to limit the exerted force according to the task requirements (e.g. allowing only small forces for scrubbing glasses, while higher forces can be exerted to clean a skillet). The main drawback of

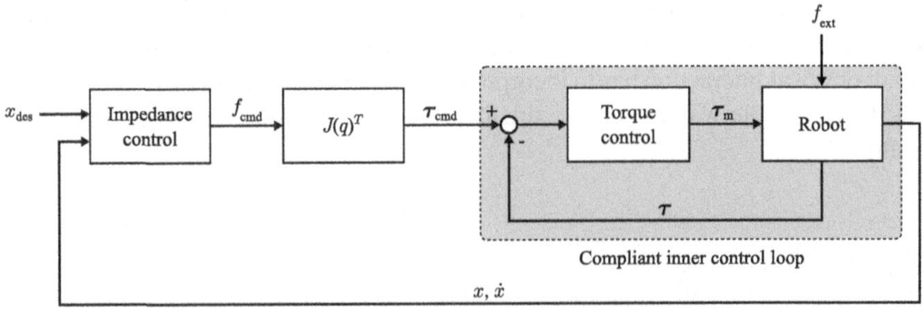

Fig. 6.2 Cartesian impedance control loop (Dietrich 2015)

this method is that the robot has to deal with the remaining steady-state error, as the impedance control law does not include an integrator. In addition, external loads have to be estimated and compensated accordingly, which is especially hard for unknown objects and objects with varying center of mass, e.g. a bottle of water. One way to circumvent this issue is to increase the stiffness. However, this also influences the task outcome. Accordingly, the parameterization has to be performed carefully, i.e. such that it does not affect the desired high-level goal of the commanded task. As a rule of thumb, only the directions of contact should be parameterized softly, if possible. For example, using a broom to clean a floor it is necessary to set a compliant behavior for the downward direction and the rotation around the brush, while it is strongly advised to set higher stiffness values for the forward direction and the upward direction in order to force the broom to travel along the desired trajectory. Otherwise, the tool might get stuck at some point and the robot fails to solve the task as indicated in Fig. 1.4. This, and some other example parameter sets are discussed in Sect. 6.3. Additionally, it might be desirable to specify different stiffness settings for different body regions or different manipulators/arms of a robot. In order to achieve this, the impedance controller has to be embedded into a whole-body control framework.

6.1.2 Hierarchical Whole-Body Control

While it is natural for humans to execute whole-body motions, it is a non-trivial task for a robot control framework to synchronize multiple manipulators and the mobile base of a (legged or wheeled) humanoid robot. In particular, a Cartesian position for one end-effector (e.g. one arm) can potentially be reached with an infinite number of joint configurations, i.e. the joint motion as well as the final joint configuration are ambiguous. While this seems to be a problem at first, it can be exploited by the control framework to solve additional control tasks simultaneously. That is, it is not only necessary to move the end-effector to the virtual equilibrium, but it is also desirable to execute this motion without traveling through singularities, without running into joint limits, and while self-collisions are avoided. As soon as multiple control tasks are executed simultaneously, it is mandatory that the more important tasks are executed with a higher priority and subordinate tasks are executed with lower priority. For example, while it is often desirable to reach the commanded goal position for the end-effector, it is even more important to avoid collisions with obstacles and surrounding humans. Moreover, while in contact, it is not trivial to determine which part of the kinematic chain shall remain in the initial position and which joints have to deviate in order to resolve the internal motions originating from the physical interaction. This becomes even more crucial if this decision influences the stability properties of the system. For this reason, a whole-body control framework has to monitor the internal state of the individual control tasks and eliminate discrepancies between them. Experience has shown that this is best incorporated by establishing a hierarchy for the individual control tasks. In the whole-body control framework implemented on Rollin' Justin, a strict task hierarchy with M priority levels is realized via *null space projections* (Nakamura et al. 1987; Siciliano and Slotine 1991), which is a standard tool in redundancy resolutions. The overall control torque can be formulated as

$$\boldsymbol{\tau} = \boldsymbol{\tau}_1 + \sum_{i=2}^{\mathrm{M}} \boldsymbol{N}_i(\boldsymbol{q})\boldsymbol{\tau}_i, \tag{6.2}$$

where the indices describe the priority levels, such that $j < k$ means that j has higher priority than k. Each of the lower level control actions $\boldsymbol{\tau}_i$ for $1, i \leq \mathrm{M}$ is projected onto the null space of all higher priority control tasks according to the respective null space projector $\boldsymbol{N}_i(\boldsymbol{q})$. Several different types of null space projectors are available with different properties. Among others, one can select between *statically consistent* (Albu-Schäffer et al. 2003) or *dynamically consistent* (Khatib 1987) null space projections and *successive* or *augmented* null space projections (Antonelli 2009). Furthermore, one specific task may require a high tracking performance (*Operational Space Formulation* (Khatib 1987)), while another tasks requires improved contact/physical interaction behavior with guaranteed stability (*Compliance Control* (Ott 2008; Dietrich et al. 2013)). However, as this exceeds the scope of this manuscript, one may refer to the overview on null space projections as given by Dietrich et al. (2015). More interesting in this context is the type of the control task and the respective parameter set. An overview is given in Table 6.1[2]. In the following, the most relevant control tasks are described based on potential fields[3].

Cartesian Impedance

Cartesian impedance (Hogan 1985; Ott 2008) is the primary control task to solve manipulation tasks in the Cartesian operational space. Cartesian impedance at an end-effectors constitutes one of the basic functionalities for any service robot. The Cartesian impedance is modeled as a mass-spring-damper system in the Cartesian directions of the TCP. This results in a compliant behavior for the robotic end-effector. The planning layer has to define the *virtual, spatial equilibrium point* $\boldsymbol{x}_{\mathrm{des}}$, the *stiffness*, the *damping*, and the maximum permissible *Cartesian forces/torques*. The control torque is

$$\boldsymbol{\tau}_{\mathrm{car}} = -\left(\frac{\partial V_{\mathrm{car}}(\boldsymbol{q}, \boldsymbol{C}_{\mathrm{car}})}{\partial \boldsymbol{q}}\right)^T - \boldsymbol{D}_{\mathrm{car}}(\boldsymbol{q}, \boldsymbol{C}_{\mathrm{car}})\dot{\boldsymbol{q}}, \tag{6.3}$$

where V_{car} denotes the spatial spring potential and $\boldsymbol{C}_{\mathrm{car}}$ describes the parameterization, e.g. the potential stiffness, the trajectories, and maximum Cartesian forces. Damping is injected by the positive definite damping matrix $\boldsymbol{D}_{\mathrm{car}}(\boldsymbol{q}, \boldsymbol{C}_{\mathrm{car}})$. Detailed information on Cartesian impedance control can be found in Ott et al. (2008).

---

[2]Although the parameterization of control tasks depends on several factors, it is not mandatory to define all parameters for every single action. In this case, general-purpose default values apply.

[3]Reactive control task for robots are most frequently defined based on artificial attractive or repulsive potential fields (Khatib 1986), where the gradient is used in the control action.

Table 6.1 Overview of the parameters for the control tasks. Note that all parameters are instantiated with general-purpose values. The overall controller parameter vector is defined as $\mathcal{C} = (\mathcal{C}_{\text{car}}, \mathcal{C}_{\text{jnt}}, \mathcal{C}_{\text{sca}}, \mathcal{C}_{\text{mes}}, \mathcal{C}_{\text{sav}}, \mathcal{C}_{\text{g}})$

Description	Value	Parameters
Cartesian impedance	$\mathcal{C}_{\text{car}}$	Stiffness (3 transl., 3 rot.), trajectories, damping, max. Cartesian wrench (3 transl., 3 rot.)
Joint impedance	$\mathcal{C}_{\text{jnt}}$	Stiffness (n joints), trajectories, joint damping
Self-collision avoidance	$\mathcal{C}_{\text{sca}}$	Stiffness (in collision direction), thresholds, emergency stop conditions, damping ratios
Avoidance of end stops	$\mathcal{C}_{\text{mes}}$	Stiffness (n joints), joint damping
Singularity avoidance	$\mathcal{C}_{\text{sav}}$	Stiffness (in singularity space), thresholds, damping (in singularity space)
Gravity compensation	$\mathcal{C}_{\text{g}}$	Loads (mass, inertia, center of mass)

Cartesian impedance is a crucial feature for resolving wiping tasks. Since it is most effective to treat all wiping actions as robot-independent problems, all assumptions on force and stiffness are made w.r.t. the involved objects and therefore in Cartesian object space (Wimböck 2013) and later transformed into the original joint coordinates.

Joint Impedance

In contrast to Cartesian impedance, joint impedance does not map the operational space to the joint space, but it is defined directly in joint space. The control torque is

$$\boldsymbol{\tau}_{\text{jnt}} = -\left(\frac{\partial V_{\text{jnt}}(\boldsymbol{q}, \mathcal{C}_{\text{jnt}})}{\partial \boldsymbol{q}}\right)^T - \boldsymbol{D}_{\text{jnt}}(\boldsymbol{q}, \mathcal{C}_{\text{jnt}})\dot{\boldsymbol{q}} \tag{6.4}$$

where V_{jnt} denotes the spring potential, the parameterization is specified by $\mathcal{C}_{\text{jnt}}$, and $\boldsymbol{D}_{\text{jnt}}(\boldsymbol{q}, \mathcal{C}_{\text{jnt}})$ is the p.d. damping matrix related to the joint impedance.

Joint impedance can be applied to arbitrary subsystems of a robot, such as an arm or the torso. This is particularly useful for bi-manual tasks, such as sweeping with a broom. In this case, the torso can be set to be stiffer than the arms to ensures a steady torso posture while external disturbances can be compensated at the same time.

Self-collision Avoidance

Especially for robots with many DOF, self-collision avoidance poses a major issue due to the large number of possible collisions between body parts. A reactive collision avoidance technique has been developed for Rollin' Justin (Dietrich et al. 2011). It is based on artificial, repulsive forces between potentially colliding links. For each collision direction, a desired mass-spring-damper relation is implemented. The control torque is

$$\boldsymbol{\tau}_{\text{sca}} = -\left(\frac{\partial V_{\text{sca}}(\boldsymbol{q}, \boldsymbol{\mathcal{C}}_{\text{sca}})}{\partial \boldsymbol{q}}\right)^T - \boldsymbol{D}_{\text{sca}}(\boldsymbol{q}, \boldsymbol{\mathcal{C}}_{\text{sca}})\dot{\boldsymbol{q}} \tag{6.5}$$

where V_{sca} denotes the repulsive potential which is parameterized in $\boldsymbol{\mathcal{C}}_{\text{sca}}$. Among others, the parameters contain the safety distance between links where the brakes are engaged (emergency stop), the potential stiffness, and the distance threshold at which the controller starts to generate repulsive forces. The damping is injected via $\boldsymbol{D}_{\text{sca}}(\boldsymbol{q}, \boldsymbol{\mathcal{C}}_{\text{sca}})$.

Self-collision avoidance may selectively be disabled for tasks where multiple manipulators work close to each other such as in the hands while washing the dishes.

Avoidance of Mechanical End Stops

Avoiding mechanical end stops is one of the standard control task in robotics. Similar to the self-collision avoidance, artificial repulsive potentials are designed around the end stops, such that this control task is only active close to them. Otherwise, it is inactive. The control torque is

$$\boldsymbol{\tau}_{\text{mes}} = -\left(\frac{\partial V_{\text{mes}}(\boldsymbol{q}, \boldsymbol{\mathcal{C}}_{\text{mes}})}{\partial \boldsymbol{q}}\right)^T - \boldsymbol{D}_{\text{mes}}(\boldsymbol{q}, \boldsymbol{\mathcal{C}}_{\text{mes}})\dot{\boldsymbol{q}} \tag{6.6}$$

where V_{mes} denotes the repulsive potential and $\boldsymbol{\mathcal{C}}_{\text{mes}}$ is responsible for the parameterization, e.g. the stiffness and the thresholds. Damping is injected via $\boldsymbol{D}_{\text{mes}}(\boldsymbol{q}, \boldsymbol{\mathcal{C}}_{\text{mes}})$.

Singularity Avoidance

With reference to Sect. 5.2.2, it is important to maintain a high manipulability throughout the entire execution of a task, in order to be able to properly react upon unexpected events or disturbances. For this reason, it is important to avoid singularities. Various methods are known in the literature. However, the representations based on the kinematic and the dynamic manipulability measure (Yoshikawa 1990) are among the most popular ones. The control torque can be described as

$$\boldsymbol{\tau}_{\text{sav}} = -\left(\frac{\partial V_{\text{sav}}(m_{\text{sav}}(\boldsymbol{q}), \boldsymbol{\mathcal{C}}_{\text{sav}})}{\partial \boldsymbol{q}}\right)^T - \boldsymbol{D}_{\text{sav}}(\boldsymbol{q}, \boldsymbol{\mathcal{C}}_{\text{sav}})\dot{\boldsymbol{q}} \tag{6.7}$$

where V_{sav} is the potential to repel from singular configurations, the manipulability measure is denoted by $m_{\text{sav}}(\boldsymbol{q})$, and $\boldsymbol{\mathcal{C}}_{\text{sav}}$ parameterizes the control task by defining the stiffness and the manipulability thresholds (Ott 2008). Again, damping is injected via $\boldsymbol{D}_{\text{sav}}(\boldsymbol{q}, \boldsymbol{\mathcal{C}}_{\text{sav}})$.

Just like mechanical end stops and self-collisions, singularities can additionally be avoided by providing roughly approximated reference trajectories to be used in combination with a joint impedance controller (6.4). This is done by calculating inverse kinematics solutions for discretized steps along the Cartesian task trajectory during the geometric reasoning step (cf. Sect. 5.3.2). Nevertheless, it is mandatory to parameterize the control tasks w.r.t. the current state of the environment, in order to react upon unforeseen disturbances such as humans in the workspace of the robot.

Gravity Compensation

Gravitational effects can be compensated by simulating a gravity model of the system online (Ott et al. 2004). The respective control torque can be written as

$$\boldsymbol{\tau}_{\mathrm{g}} = \left(\frac{\partial V_{\mathrm{g}}(\boldsymbol{q}, \boldsymbol{C}_{\mathrm{g}})}{\partial \boldsymbol{q}} \right)^T , \tag{6.8}$$

where V_{g} denotes potential of the gravity vector parameterized by $\boldsymbol{C}_{\mathrm{g}}$, which contains information about the mass, the inertia, and the center of mass for all links of the robot and manipulated objects.

Platform Control

Nonholonomic, mobile platforms are mostly controlled via kinematic controllers. The robot follows a commanded trajectory while implicitly complying with the rolling constraints. In the case of Rollin' Justin, such a control law has been implemented by Giordano et al. (2009). The controller expects a trajectory in the plane of motion, i.e. in one rotational and two translational directions.

Mobility is inevitable to solve manipulation tasks that occupy wide areas (e.g. cleaning the floor or wiping large windows). Moreover, combining the mobile base with additional joints is required to execute whole-body motions. Whether to move only a subset of joints, arms, and torso, or to include the mobile base for solving a given task has to be defined during the reasoning process.

Altogether the control tasks span a large parameter space as one can see in Table 6.1. The control framework alone cannot parameterize this space meaningfully, semantically speaking, i.e. by incorporating the task-specific requirements. This is due to the lack of semantic information on the control level. Accordingly, this problem has to be forwarded to the reasoning layer, which is the subject of the following section.

6.2 Controller Parameterization

Free space motions that are executed with an impedance controller can be modeled in such a way that the motion of the robot is relatively accurate with a marginal tracking error. In contact, the motion deviates naturally. The deviation of the motion results in a force at the end-effector. This causality can be exploited in the design of compliant manipulation tasks. In order to prevent damage to the robot and the environment, it is strongly recommended to limit the exerted forces to a certain maximum. The

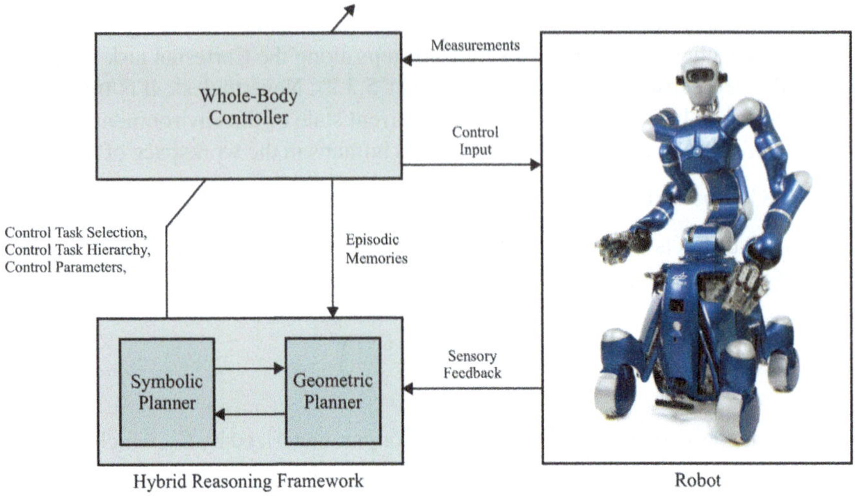

Fig. 6.3 Overview of the cognition-enabled controller concept. The whole-body controller is parameterized by the reasoning introduced in Chap. 5. In return, the controller provides episodic memories to interpret the task outcome to be detailed in Chap. 7

parameterization of these values cannot be done by the control level itself as it has no knowledge on the high-level task context. Only by integrating high-level task reasoning with low-level impedance control an appropriate parameterization can be achieved. An overview of this cognition-enabled controller concept is given in Fig. 6.3.

6.2.1 Trajectory Design in Contact

Symbolically, the effects of an action are meant to be described according to the classification of compliant manipulation tasks conducted in Chap. 3. The symbolic planner selects the correct action based on the current symbolic world state, i.e. the available tools and the given task to solve. In the example of wiping actions, this means that the action involves the tool-medium-surface tuple. The geometric parameterization as well as the controller parameters are selected to meet the requirements of the particular instance of the task. This parameterization is performed during the reasoning procedure. Depending on the parameter type a part of the object knowledge is responsible for the parameterization. The hierarchy can be set during the execution of the Action Template. An example parameterization for the accumulation of shards using a broom is outlined in Fig. 6.4. Here, the self-collision avoidance is of top priority, which is often the case when multiple manipulators interact simultaneously with the environment. Second, the Cartesian impedance is parameterized. The Cartesian task trajectory specifies the virtual, spatial equilibrium point over time, where the

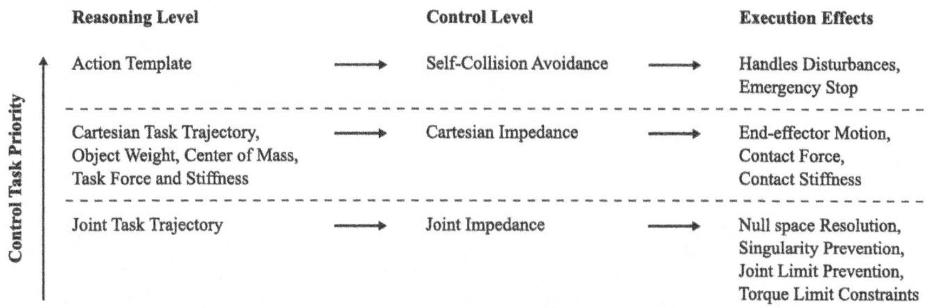

Fig. 6.4 The conceptual parameterization of the control level. The objects that participate in the action span the parameter set. The respective object elements that parameterize the control level are listed on the left, the influenced controller task is listed in the center, and the execution effects that are mandatory to fulfill the desired task outcome are listed on the right

information about the object mass and the center of mass have to be incorporated as well. The desired contact behavior is regulated based on the forces and torques in the operational space. Additionally, a joint trajectory may be provided to generate a predictable whole-body motion that avoids singularities and joint limits.

On the geometric level the task is described according to geometric relations between the involved objects. However, the compliant behavior of the real robot is difficult to model; even when excluding possible elastic contact behavior between tool and surface. To solve this issue human task knowledge is exploited by integrating the control level into the reasoning step. This is done by defining the Cartesian trajectory for the TCP $\boldsymbol{x}_{\text{tool}}$ w.r.t. the virtual, spatial equilibrium point $\boldsymbol{x}_{\text{des}}$ for the end-effector. The trajectory is computed w.r.t. the involved objects and their properties, e.g. the mug radius defines the circular trajectory for the sponge. In this example, only translational motions are commanded.

At control level, the compliance parameters are defined. The rotations of the tool are introduced by defining the Cartesian object stiffness and the maximum Cartesian contact force. Deviations on contact are neglected during planning time. The resulting Cartesian control torque $\boldsymbol{\tau}_{\text{car}}$ according to (6.3) enables a compliant robot to adapt itself to the curvature of the target surface resulting in a deliberate deviation of the end-effector from the commanded Cartesian trajectory

$$\boldsymbol{x}_{\text{act}} = \boldsymbol{x}_{\text{des}} + \boldsymbol{x}_{\text{dev}}\left(\boldsymbol{q}, \boldsymbol{C}, \boldsymbol{f}_{\text{ext}}\right). \tag{6.9}$$

Herein $\boldsymbol{x}_{\text{act}}(\boldsymbol{q})$ describes the forward kinematics based on the link side measurements (see Fig. 6.5). It contains both the translation $\boldsymbol{p}_{\text{act}}(\boldsymbol{q})$ of the TCP and its orientation, depending on the chosen representation (e.g. via Euler angles). The term $\boldsymbol{x}_{\text{act}}(\boldsymbol{q})$ can be represented as the combination of the commanded TCP position/orientation $\boldsymbol{x}_{\text{des}}$ and the deviation $\boldsymbol{x}_{\text{dev}}$ between this commanded and the actual TCP position/orientation. The deviation is subject to the parameterization $\boldsymbol{C}$ of the whole-body controller and the external forces/torques $\boldsymbol{f}_{\text{ext}}$.

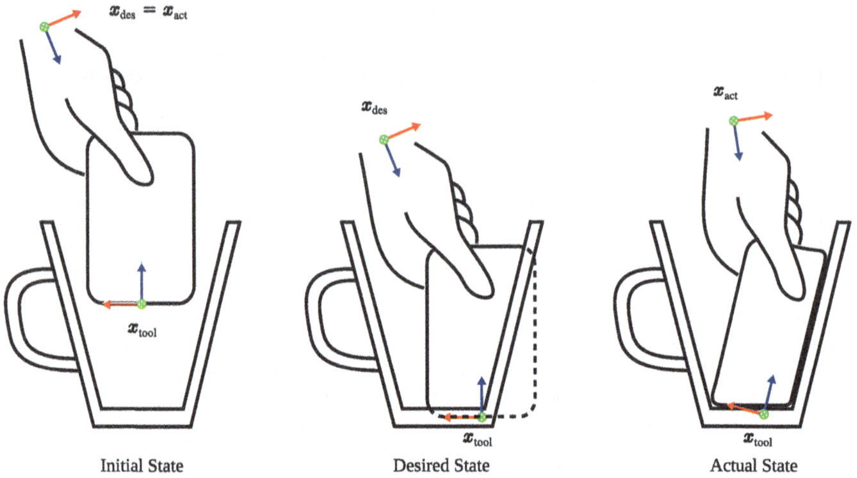

Fig. 6.5 The compliant control strategy can be exploited to simplify the planning of in-contact motions. The commanded state for the tool x_{tool}, resulting from the virtual, spatial equilibrium x_{des} of the right end-effector, is geometrically unfeasible. In contact, the parameterization of the robot forces the correct alignment of the tool (including deformations) to solve the task (see x_{act})

The control level alone cannot guarantee a successful task execution by only following the Cartesian virtual equilibrium, since the applied local control methods are vulnerable to get stuck in local minima. Furthermore, collisions with the obstacles other than the robot cannot be prevented that way. Therefore, a roughly approximated reference trajectory is provided during the reasoning step by computing discrete inverse kinematics solutions along the task trajectory (cf. Sect. 5.3.2). During run-time, the reference trajectory is interpolated to provide an input for the joint impedance controller based on (6.4). The joint impedance control action is projected onto the null space of the Cartesian impedance resulting in an overall whole-body motion of the robot, maintaining contact and free of local minima. Additional control torques for self-collision avoidance τ_{sca} (6.5), avoidance of mechanical end stops τ_{mes} (6.6), and singularity avoidance τ_{sav} (6.7) are computed with lower control task priority to react on unforeseen events.

6.2.2 Controlling the Contact Force

The discussed control concepts are integrate into the compliant whole-body impedance framework introduced by Dietrich et al. (2012). The soft physical contact behavior is a natural outcome of this framework. It does not apply a force-control strategy, but rather exploits the compliant impedance behavior to realize compliant motions into the target. At the example of wiping tasks, this results in a motion that is aligned with the target surface. The controller force at the end-effector is saturated to match the desired contact force.

The controller force f_{cmd} results from the Cartesian impedance controller, defined by the virtual elastic potential V and the damping matrix $\boldsymbol{D}$. It is saturated and transformed into the joint space by the transposed Jacobian, such that

$$\tau = -\boldsymbol{J}^T \underbrace{\boldsymbol{\mathcal{S}}\left(\left(\frac{\partial V}{\partial \boldsymbol{x}}\right)^T + \boldsymbol{D}\dot{\boldsymbol{x}}\right)}_{\boldsymbol{f}_{\mathrm{cmd}}}, \tag{6.10}$$

where $\boldsymbol{\mathcal{S}}$ is the saturation function for the two force terms $\boldsymbol{f}_{\mathrm{spring}} = (\partial V/\partial \boldsymbol{x})^T$ plus $\boldsymbol{f}_{\mathrm{damper}} = \boldsymbol{D}\dot{\boldsymbol{x}}$. The external contact force $\boldsymbol{f}_{\mathrm{ext}}$ counteracts the controller force $\boldsymbol{f}_{\mathrm{cmd}}$ under the assumptions of a static case with negligible model uncertainty. The transformation of the TCP is thereby defined as

$$\boldsymbol{H}_{\mathrm{tcp},i} = \boldsymbol{H}_{\mathrm{act},i} \cdot \boldsymbol{H}_{\mathrm{grasp}}^{-1}, \tag{6.11}$$

where the homogeneous transformation matrices $\boldsymbol{H}_*$ correspond to the task space coordinates $\boldsymbol{x}_*$. The grasp transformation matrix $\boldsymbol{H}_{\mathrm{grasp}}$ is assumed to be constant during the task execution. That is, the tool is firmly grasped by the robot such that it does not significantly move in the hand of the robot. This way, the mechanical deformations of the tool can be neglected, which corresponds to the illustration in Fig. 6.5[4].

As a result, the manipulated tool aligns naturally with the target if the Cartesian stiffness is carefully parameterized. The forces are limited such that the task-specific requirements can be met. In conclusion, impedance control with saturated Cartesian forces is highly recommended for the execution of wiping motions.

6.2.3 Discussion

The control strategy for a particular process model can be represented by a distinct Action Template. The process model involves the symbolical meaning and the geometric execution of the control strategy. The required sub-symbolic parameters such as task frames or regions of interest are provided by the involved objects. Additionally, the actual controller parameterization (including the maximum permitted contact force and the end-effector stiffness) is inherent. Preprocessed by a human programmer, this task-related knowledge is used to encode the desired compliant control behavior for the wiping action according to its classification. Such a behav-

[4]Alternatively, one could improve the design of the trajectory, such that the tool follows the curvature of the target surface accurately. By additionally defining a bias force, the controller would converge toward the desired contact.

ior can hardly be generalized by using local control strategies only. Moreover, there is no geometric planner known which could possibly simulate such a behavior sufficiently stable in a generalized way. Therefore, the control level and the reasoning level have to act jointly.

Using Action Templates to program and parameterize robotic manipulation actions is beneficial since the programmer is motivated to define the process model based on predicted physical behavior and human task knowledge, which is otherwise hardly transferred to the control level of a robot. In the following section, this approach is demonstrated for three whole-body wiping tasks executed on Rollin' Justin.

6.3 Execution of Compliant Wiping Motions

This section evaluates the control strategy and the parameterization of the whole-body impedance controller in practice. Three distinct everyday household chores are planned and executed. The experiments are designed as whole-body manipulation tasks inspired by natural everyday household chores in human environments. The first experiment addresses the mug scrubbing task already introduced in the previous section. The second experiment is a window wiping task in which the robot has to clean a large window with a one-handed window wiper. The third experiment is a sweeping task where shards of broken dishes have to be collected by bi-manually handling a broom. Each experiment demonstrates one particular wiping class. Due to the classification and the resulting varying process models the tasks have to be parameterized differently. This section describes directions and rotations in robot coordinates (positive x-axis forward, positive y-axis left, positive z-axis up), while the Cartesian force and stiffness parameters are defined in the local TCP coordinate systems of the involved tools as illustrated in Table 6.2.

Table 6.2 Table of tool-specific parameters. The Cartesian force and the Cartesian stiffness parameters are given w.r.t. the visualized frames

	processinlinefigure	processinlinefigure	processinlinefigure
Action template	_sponge.scrub	_wiper.skim	_broom.collect
Actuators	both arms, torso	one arm, torso, base	both arms, torso, base
Control hierarchy	$\mathcal{T}_{\text{car}}, \mathcal{T}_{\text{jnt}}, \mathcal{T}_{\text{mes}}, \mathcal{T}_{\text{sav}}$	$\mathcal{T}_{\text{car}}, \mathcal{T}_{\text{sca}}, \mathcal{T}_{\text{sav}}, \mathcal{T}_{\text{jnt}}, \mathcal{T}_{\text{mes}}$	$\mathcal{T}_{\text{car}}, \mathcal{T}_{\text{sca}}, \mathcal{T}_{\text{jnt}}, \mathcal{T}_{\text{mes}}, \mathcal{T}_{\text{sav}}$
Trans. stiffness x, y, z (N/m)	400, 400, 800	100, 500, 1000	1000, 500, 300
Rot. Stiffness $\theta_x, \theta_y, \theta_z$ (Nm/rad)	30, 30, 60	500, 10, 10	200, 10, 500
Cart. force limits x, y, z (N)	$\pm 20, \pm 20, \pm 20$	$-\infty/+10, \pm\infty, \pm\infty$	$\pm\infty, \pm\infty, -10/+\infty$

6.3.1 Scrubbing a Mug with a Sponge

The first experiment is based on the conceptual example described in Fig. 6.5, i.e. scrubbing a mug with a sponge. The goal of this task is to clean the inner curvature of the mug. This action is classified as *scrubbing*. The tool is the sponge and the target surface is the inner region of the mug. The surface area can actively be positioned if the mug is grasped by the robot. There are two media involved. The first medium is the dried-up staining in the mug. The second medium is the detergent absorbed by the sponge. The symbolic representation is defined in the Action Template as follows:

Listing 6.1 Symbolic action representation of the mug scrubbing experiment.

```
(action _sponge.scrub:
    :parameters (?t - _sponge ?s - _dish
                 ?m1 - _medium ?m2 - _medium
                 ?a1 - _manipulator ?a2 - _manipulator)
    :precondition (and (absorbed ?t ?m1) (adhesive ?s ?m2)
                       (picked ?t ?a1) (picked ?s ?a2))
    :effect (and (not(adhesive ?s ?m2)) (scrubbed ?s ?m2))
)
```

From a semantic point of view, the mug has to be cleaned, which requires to combine detergent (previously absorbed by the sponge) with the coffee leftovers in the coffee mug by executing a scrubbing action. As a precondition, the mug and the sponge have to be grasped simultaneously by the robot, which requires two distinct manipulators. The anticipated effects are that the dirt is not anymore adhesive to the mug but scrubbed, such that it can be washed out. Please note that the effects are not validated visually within this example. Suitable methods are introduced in Chap. 7 to infer the effect of wiping motions based on haptic perception.

In terms of reasoning, this task corresponds to the example introduced earlier in Sect. 5.1. As a reminder, the mug needs to be picked up upside-down to clean the inner curvature of it, which is not feasible by only executing one single pick action. The symbolic representation of the Action Template requires the mug and the sponge to be picked as precondition. Only after backtracking on the geometric level and on the symbolic level by cutting the symbolic domain the symbolic planner finds the correct solution by scheduling a handover action as deduced in Sect. 5.1.2:

Listing 6.2 Symbolic transition for the mug scrubbing experiment.

```
_object.pick mug left_arm table,
_object.handover mug left_arm right_arm,
_object.pick sponge left_arm table,
_sponge.scrub sponge left_arm mug
```

The geometric representation of the `_sponge.scrub` action defines a circle which is twice as big as the diameter of the mug. The sponge aligns with the curvature of the mug and the TCP deviates significantly (plotted as green solid line in Fig. 6.6)

Fig. 6.6 The humanoid robot Rollin' Justin scrubbing a mug with a sponge

from the spatial, virtual equilibrium point (plotted as blue solid line in Fig. 6.6). This results in forces on the medium which in turn results in the removal of the dirt. The translational and rotational stiffness for the left hand (holding the sponge) are set to be low along/about the lateral axis of the sponge as described in Table 6.2. The right hand (holding the mug) is commanded to be stiff for all rotations and translations. With this parameterization the robot is able to solve the given wiping task. Figure 6.7 shows the resulting measured trajectory of the manipulator. During contact, the force limitations in Table 6.2 apply.

Regarding whole-body control, an active torso and base are not mandatory to clean the dishes. However, by treating the action equally to a whole-body task, the torso and the base can be used to compensate for external disturbances and avoid obstacles while simultaneously remaining over the sink with potentially wet dishes. Self-collision avoidance is disabled during this task to enable both manipulators to approach each other. The control task hierarchy is $\boldsymbol{\tau}_{\text{car}}, \boldsymbol{\tau}_{\text{jnt}}, \boldsymbol{\tau}_{\text{mes}}, \boldsymbol{\tau}_{\text{sav}}$.

6.3.2 Skimming a Window with a Window Wiper

Window cleaning is a diverse *skimming* task. It can be solved with several tools, one-handed or bi-manually. It is applicable to many window types such as squared, round, or curved ones as outlined in Sect. 5.1.1. Based on the size of the window

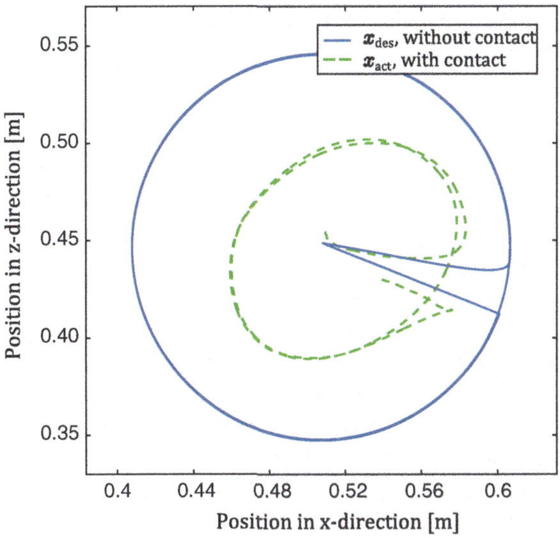

Fig. 6.7 Desired and actual position of the left hand during scrubbing

the whole body might be required to reach the entire window pane. An individual parameterization is mandatory w.r.t. individual settings.

The problem definition of the window wiping scenario at hand has already been introduced with the reachability analysis conducted in Sect. 5.2.2. It is emphasized that the window pane is too large to reach it without the aid of the torso and the mobile base. This makes the task an intersecting use-case to illustrate the parameterization of the control task hierarchy, which is $\mathcal{T}_{car}, \mathcal{T}_{sca}, \mathcal{T}_{sav}, \mathcal{T}_{jnt}, \mathcal{T}_{mes}$. Accordingly, the Cartesian task motion $\mathcal{T}_{car}$ has the highest priority. This is most frequent the case as the end-effector motion is crucial to accomplish a manipulation task. Second, the self-collision avoidance $\mathcal{T}_{sca}$ is added. This is due to the fact that the action is performed close to the head of the robot, which has to be avoided at all cost as it carries the stereo camera system of the robot. Next, it is beneficial to set the control task priority for $\mathcal{T}_{sav}$ to prevent singularities, especially in the corners of the window. This sub-goal goes hand in hand with the joint impedance $\mathcal{T}_{jnt}$. It is fed with a preplanned joint motion that avoids collisions with the environment. At last, the avoidance of mechanical end stops is added which is necessary as the joint deviations may get too high in case of contact with the window pane. Altogether, the robot is enabled to execute the wiping task sufficiently accurately and safely. In the example at hand, the frame of the window is initially localized visually, such that the robot is able to approach the window autonomously.

Listing 6.3 Symbolic action representation for the window skimming experiment.

```
(action _wiper.skim:
  :parameters (?t - _wiper ?s - _window
               ?m - _detergent ?a1 - _manipulator)
  :precondition (and (picked ?t ?a1)
                     (applied ?s ?m))
  :effect (and (skimmed ?s ?m) (not (applied ?s ?m)))
)
```

The task is parameterized according to prior task knowledge deposited w.r.t. the involved objects. From a symbolical planning point of view the detergent (applied by a human co-operator) has to be removed from the surface area by skimming it away. As a result, the medium, i.e. the detergent is no longer applied to the surface. The symbolic representation is defined as seen in Listing 6.3. As precondition, the window wiper has to be grasped by the robot, while a detergent is applied to the surface. The resulting symbolic transition consists of a pick and a skim action. Again, the result of the skim action is not visually validated.

Listing 6.4 Symbolic transition for the window skimming experiment.

```
_object.pick wiper table right_arm,
_wiper.skim window right_arm
```

The skim Action Template describes the geometric wiping task as SDG (cf. Sect. 5.3). The ROI is defined in the data storage of the mock-up window. Together with the dimension of the wiper blade the Cartesian task trajectory can be computed as shown in Fig. 6.9. The TCP of the wiper is defined in the center of the blade and is used to describe the tool alignment. The object stiffness parameterization for the window wiper is listed in Table 6.2. Special attention is given to the rotational stiffness, which refines the alignment by introducing freedom about the y-axis and the z-axis of the object TCP. This way a compliant behavior of the blade can be achieved, leading to a more robust task execution. As already described in Fig. 6.5, the wiper aligns near optimal with the target surface. A rather soft parameterization allows to overcome local errors as well as external disturbances. The Cartesian force is predefined by the window wiper and may vary from tool to tool. In this case the robot exerts a maximum Cartesian force of $f_x^{\max} = 10\,\text{N}$.

One of the main difficulties with the manipulation of windows is the visual perception. The transparent glass surface offers little visual feedback, such that the window pane can mainly be estimated by localizing the window frame. However, even with relatively inaccurate localization the robot is able to execute the commanded task. The compliant contact behavior is the key to execute the task successfully in imperfect conditions. This becomes evident when comparing the desired trajectory $\boldsymbol{x}_{\text{des}}$ with the actual trajectory $\boldsymbol{x}_{\text{act}}$. In particular, the robot is able to re-align the window wiper with the window pane, even though it collided with the window frame as it is observed in the first row in the upper left of Fig. 6.8. The issue of imperfect localization is further investigated in Chap. 7.

Fig. 6.8 This plot compares the spatial, virtual equilibrium point $\boldsymbol{x}_{\text{des}}$ with the actual position of the right hand $\boldsymbol{x}_{\text{act}}$ during the window wiping task. The three axes correspond to the three dimensions of the window (similar to robot coordinates), where z is the height and y is the width. The deviation along the x-axis corresponds to the contact of the wiper with the window pane. Note that the transit paths (connecting two contact situations) are removed for clarity

6.3.3 Collecting Shards with a Broom

The third task requires a *collect* action where a broom is utilized to collect shards of a broken mug. This task is designed to be similar to the motivational task of collecting leaves with a rake which was outlined in Chap. 1. In the introduction, this task is chosen as running example that should guide the reader through the manuscript. Eventually, this section emphasizes how this manuscript contributes to this goal by tackling and indoor equivalent of this task. Similar to raking leaves, collecting shards with a broom requires both arms to handle the broom while the mobile base and the torso support the motion to create an overall wiping trajectory along a larger area. Both arms, the torso, and the mobile base act jointly. The deep integration of AI reasoning methods with advanced control mechanisms for humanoid robots is mandatory. Similar to the window cleaning task, the broom needs to be picked before it can be used. Since the broom has to be handled bi-manually, the mass of the tool has to be carried by both manipulators. The pick Action Template is therefore specialized by the `_broom` object class. The symbolic representation of the collect Action Template is defined as follows:

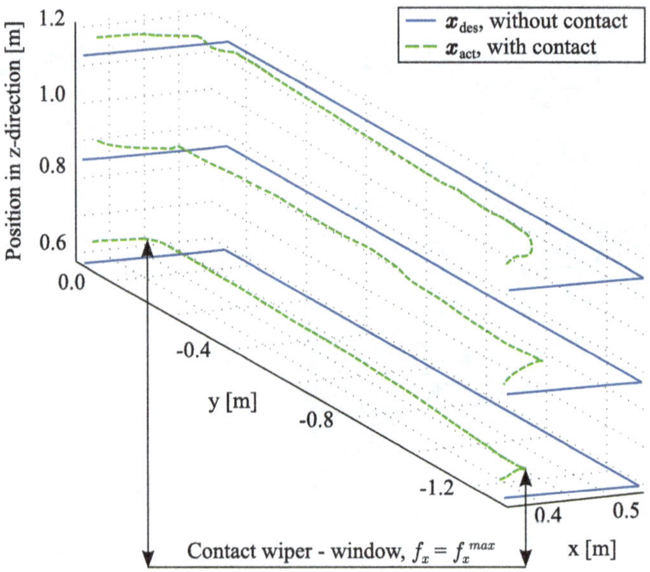

Fig. 6.9 Rollin' Justin wiping a window. Based on the ROI (red) and the wiper width the Cartesian task trajectory (blue, dashed arrows) is computed. The task has been successfully demonstrated to a public audience at the Hanover Fair 2013

Listing 6.5 Symbolic action representation for the collect experiment.

```
(action _broom.collect:
  :parameters (?t - _broom ?s - _floor ?m - _dish
               ?a1 - _manipulator ?a2 - _manipulator)
  :precondition (and (picked ?t ?a1) (picked ?t ?a2)
                     (broken ?m))
  :effect (and (collected ?s ?m))
)
```

This example illustrates the use of a medium other than small particles or liquids. As defined in the symbolic precondition section, the fragments of the mug can only be collected with the broom if it is marked as broken. In this form the mug constitutes the medium for the wiping task. The ROI spanned by the shards was however predefined. Similarly, the result of the wiping action is not visually validated. In case of the desired symbolic goal state collected floor mug, the symbolic planner yields the following symbolic transition:

Listing 6.6 Symbolic transition for the collect experiment.

```
_broom.pick broom table right_arm left_arm,
_broom.collect broom mug right_arm left_arm
```

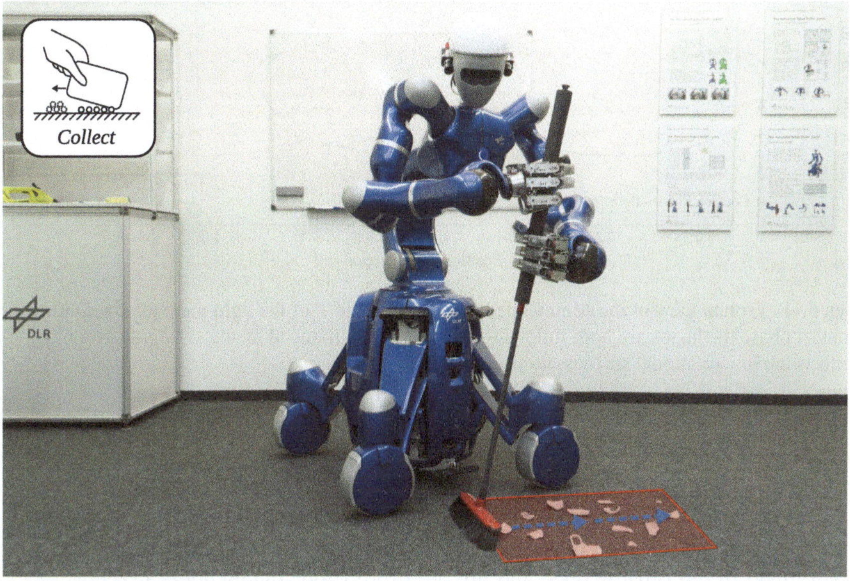

Fig. 6.10 Rollin' Justin collecting shards of a broken mug. The ROI (red) is defined by the distribution of the shards which is here localized manually. Along this region, a cyclic trajectory has to be executed (blue, dashed arrow)

The experiment is shown in Fig. 6.10. In practice, the shards would define the ROI in this particular collecting task which spans the target area, i.e. the search space for the SDG. While the visualized scene shows a limited distribution of the shards, it may also be the case that shards are spread over a wider area, such that the robot has to be repositioned. However, the Cartesian task trajectory is predefined here in order to bring the control strategy into focus. Accordingly, the Cartesian task trajectory is designed as a simple ellipse intersecting the floor plus a linear lateral offset. The elliptic motion is to be executed by the arms and the lateral motion by the mobile base. This way the contact during the trajectory execution is slightly overlapped. The trajectory can be imagined as a loop (see Fig. 6.11). With this trajectory alone the robot is not able to collect the shards with the broom. Only after setting the correct controller parameters the task can be solved. The task hierarchy is $\boldsymbol{\tau}_{\text{car}}, \boldsymbol{\tau}_{\text{sca}}, \boldsymbol{\tau}_{\text{jnt}}, \boldsymbol{\tau}_{\text{mes}}, \boldsymbol{\tau}_{\text{sav}}$, which is almost identical to the window wiping task. However, the singularity avoidance is of least priority as singular joint configurations are unlikely for this task. The desired motion of the tool involves a tilted TCP, wiping along the floor (see Table 6.2). This requires a low rotational stiffness along the brush attachment (y-axis) and a low translational stiffness along the shaft (z-axis). The force against the ground (z-axis) is limited to $f_z^{\min} = -10\,\text{N}$. The desired Cartesian object stiffness for the broom is listed in Table 6.2. The coupling of stiffness and damping for bi-manual tool handling can be parameterized based on Wimböck (2013); Florek-Jasinska et al. (2014). In addition to the Cartesian stiffness, the joint stiffness for

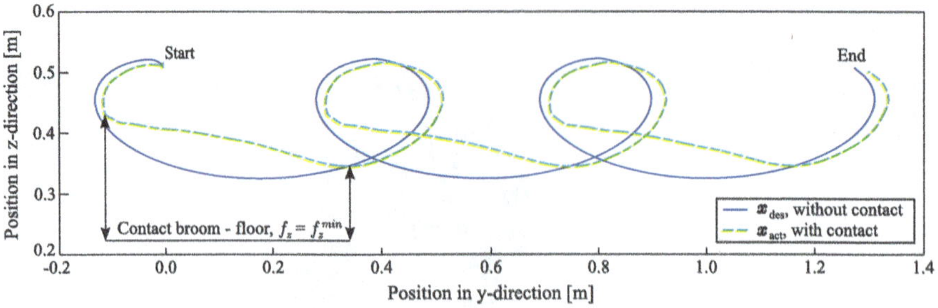

Fig. 6.11 Frontal view of the commanded and measured TCP of the right manipulator. During the contact phase the hands are both following a tilt motion introduced by the sweeping broom which reflects the low rotational stiffness around the brush

the torso is defined to be twice the default value, such that the main compliance is anticipated by the arms.

This experiment shows how whole-body manipulation of closed kinematic chains can be integrated into a hybrid reasoning framework. All relevant assumptions have been made in the object space and are therefore robot-independent (except for the torso parameterization, which is exemplary, not mandatory). The resulting behavior solves the commanded sweeping task. Eventually all shards are collected in one area. The actual measured trajectory of the right manipulator in Fig. 6.11 shows that it is not straightforward to define the correct task motion by programming it in detail. The local adaption on control level paired with globally applied human task knowledge is a well-suited method to solve this issue. Moreover, the same Action Template would be suitable to solve the task of collecting leaves as it is sketched in Fig. 1.1. It would be only required to change the parameters for the particular tool-medium-surface tuple, i.e. the combination of the parameters for the rake, the leaves, and the cobblestone. The tool motion can be calculated online by means of the SDG algorithm according to the visual perception of the leave distribution (represented as particle set $\mathcal{P}$).

6.3.4 Discussion

The experiments show that the concept of Intelligent Physical Compliance is suitable to represent, plan, and execute wiping tasks of different kind that require compliant whole-body coordination. For all tasks the same controller was utilized. It is shown that various components can be addressed using the proposed reasoning framework to solve even complex whole-body manipulation tasks rather than only pick-and-place tasks. All information needed to describe the tasks is stored within the context of the involved objects. The task reasoning and controller parameterization is autonomously performed by the robot with the use of the provided Action Templates.

The main advantage of the approach presented in this section is the relative ease with which a manipulation task can be defined in order to act correctly on the control level. A programmer has to develop a single Action Template only, utilizing prior task knowledge including the task trajectory, Cartesian object stiffness, and Cartesian force for the controller parameterization. However, it is not straightforward to select the correct parameters. Unfortunately, no detailed studies are available on exerted force and stiffness during manipulation tasks, neither for robots nor for humans. This lack can be mainly traced back to impracticable measuring procedures during tool usage. Nevertheless, Cartesian tool trajectories can be tracked. Doing so it would be possible to automatically learn the correct force, torque, and stiffness parameterization with suitable machine learning strategies, providing the desired Cartesian trajectory by human demonstration. Similarly, a robot could learn parameter settings from related tasks that were manually parametrized by a human expert.

One major issue discovered during the research presented in this chapter, is that the effects of wiping motions are typically hard to estimate. However, quality estimation is a major requirement for service robots that perform cleaning tasks. While it might be possible to detect shards of a broken mug quite well, it is impossible to identify dust or small dirt particles on a carpet. For this reason, Chap. 7 introduces a model based approach to estimate the task performance of wiping actions from haptic feedback.

6.4 Related Work

The works of Yamamoto and Yun (1992), Tan et al. (2003), Moro et al. (2013), and Dietrich et al. (2012) show how mobile manipulation is successfully tackled at the control level. The *Operational Space Formulation* by Khatib (1987) is among the most popular methods to implement force control in a reduced space (Sentis and Khatib 2005; Sadeghian et al. 2014), e.g. in the Cartesian coordinates of the end-effector. All of these works are limited to the control aspect of the problem. However, parameterizing controllers for different tasks without any high-level task context and without the knowledge about the involved objects is not possible.

Some research groups have investigated the control problem for wiping motions in contact in more detail. Ortenzi et al. (2014) propose to exploit the environment contact constraints of wiping tasks in the operational space, to decouple the motion of the robot from the applied force. As a result, the interaction requires less torque while the position accuracy is comparable to other approaches. Schindlbeck and Haddadin (2015) utilize task-energy tanks to react safely upon contact loss. This issue is critical in the absence of visual perception (or poor visual perception as it is observed for poor lighting conditions) as it is detailed in Chap. 7.

The combination of AI-based reasoning methods and control theory for whole-body manipulation, is a rare research topic. Some of the most notable approaches so far utilize constraint-based controller parameterization. Bartels et al. (2013) ground symbolic actions by the use of a *constraint-based movement description language* based on geometric features, such as points, lines, and planes to be interpreted by the control level. In particular, the authors argue that this representation enables a robot to execute manipulation tasks in a feedback-driven manner and allows for high-level error monitoring. The representation can be used for manual or autonomous generation of motion commands and can be extended toward compliant interaction. Vanthienen et al. (2013) propose an alternative way of describing describe wiping tasks as a set of constraints with the *iTaSC* framework. This approach does integrate the parameterization of soft contacts. As the iTaSC framework is a widely used tool, it supports several robotic platforms and is therefore a valuable tool toward the development of universal service robots by the research community.

Most recently, Hazara and Kyrki (2016) proposed an approach to imitate wiping motions from human demonstration. That is, the article formulates a solution to the proposal raised in the discussion at the end of Sect. 6.3.4. The approach integrates impedance control for the execution of wiping motions with state-of-the-art machine learning methods to avoid complex planning methods, and automatically recall and enhance the previously taught motions instead. In particular, the authors utilize *Dynamic Motion Primitives* and apply the *Path Integral* (PI^2) algorithm to update the imitated force policy. As a result, the deployed light-weight robot is able to successfully execute a wood grinding task repeatedly with different wooden planks. Similarly to the experiment conducted in this Sect., Silvério et al. (2015) teach a bi-manual robot the task of sweeping with a broom. The authors assume constant stiffness and damping gain matrices. The contact force is mostly neglected. They utilize a task-parameterized formulation of a Gaussian mixture model to learn the relation of the end-effector positions and orientations simultaneously. This way, the robot is able to adapt to sweeping motions with different orientations, without violating the constraints introduced by the closed kinematic chain.

6.5 Summary

Compliant contact behavior is the hallmark of modern light-weight robots. However, up to now, no robots have fully exhausted the possibilities that accompany this feature. The reason for this is that it is not straightforward to program a robot to execute compliant in-contact tasks. This is especially problematic as these kind of tasks are most frequently observed in domestic and industrial settings. While the control strategies for compliant manipulation are under development since the first torque-controlled robots emerged, the generic development of goal-oriented actions that require soft contact have been neglected for quite some time. As a result, it is still an open question how these sophisticated control strategies can be efficiently integrated and parameterized such that they are applicable as standard tools while still being able to match the requirements for a specific task. In this chapter, an approach was formulated to integrate a compliant whole-body controller into an AI-based hybrid reasoning framework to overcome this issue. The reasoning framework set up the controller based on context-specific information provided by the objects that participate in the task execution. The control tasks (e.g. Cartesian impedance, joint impedance, self-collision avoidance, etc.) were selected, hierarchically prioritized, and individually parameterized. This way, one controller can be applied to manifold tasks. In this chapter, three experiments were conducted, namely scrubbing mug with a sponge, skimming detergent off a window, and collecting shards with a broom. The developed approach enabled the robot Rollin' Justin to perform the three tasks successfully. As such, the chosen approach constitutes a suitable method for the execution of compliant manipulation tasks. The following chapter is directly linked to this statement by introducing an approach to estimate the performance of robotic wiping motions qualitatively.

References

Albu-Schäffer, Alin, Christian Ott, Udo Frese, and Gerd Hirzinger. 2003. Cartesian Impedance Control of Redundant Robots: Recent Results with the DLR-Light-Weight-Arms. In *Proceedings of the IEEE International Conference on Robotics and Automation (ICRA)*, pp. 3704–3709.

Antonelli, Gianluca. 2009. Stability Analysis for Prioritized Closed-loop Inverse Kinematic Algorithms for Redundant Robotic Systems. *IEEE Transactions on Robotics* 25 (5): 985–994.

Bartels, Georg, Ingo Kresse, and Michael Beetz. 2013. Constraint-Based Movement Representation Grounded in Geometric Features. In *Proceedings of the IEEE/RAS International Conference on Humanoid Robots (ICHR)*, pp. 547–554.

Dietrich, Alexander. 2015. *Whole-Body Impedance Control of Wheeled Humanoid Robots*. PhD thesis, Technische Universität München (TUM).

Dietrich, Alexander, Christian Ott, and Alin Albu-Schäffer. 2013. Multi-objective Compliance Control of Redundant Manipulators: Hierarchy, Control, and Stability. In *In Proceedings of the IEEE/RSJ International Conference on Intelligent Robots and Systems (IROS)*, pp. 3043–3050.

Dietrich, Alexander, Christian Ott, and Alin Albu-Schäffer. 2015. An Overview of Null Space Projections for Redundant, Torque-controlled Robots. *International Journal of Robotics Research* 34 (11): 1385–1400. https://doi.org/10.1177/0278364914566516.

Dietrich, Alexander, Thomas Wimböck, Holger Täubig, Alin Albu-Schäffer, and Gerd Hirzinger. 2011. Extensions to Reactive Self-collision Avoidance for Torque and Position Controlled Humanoids. In *Proceedings of the IEEE International Conference on Robotics and Automation (ICRA)*, pp. 3455–3462.

Dietrich, Alexander, Thomas Wimböck, Alin Albu-Schäffer, and Gerd Hirzinger. 2012. Reactive Whole-body Control: Dynamic Mobile Manipulation Using a Large Number of Actuated Degrees of Freedom. *IEEE Robotics & Automation Magazine* 19 (2): 20–33.

Friedl, Werner, Hannes Höppner, Florian Petit, and Gerd Hirzinger. 2011. Wrist and Forearm Rotation of the DLR Hand Arm System: Mechanical Design, Shape Analysis and Experimental Validation. In *Proceedings of the IEEE/RSJ International Conference on Intelligent Robots and Systems (IROS)*, pp. 1836–1842. IEEE.

Florek-Jasinska, Monika, Thomas Wimböck, and Christian Ott. 2014. Humanoid Compliant Whole Arm Dexterous Manipulation: Control Design and Experiments. In *Proceedings of the IEEE/RSJ International Conference on Intelligent Robots and Systems (IROS)*, pp. 1616–1621.

Grebenstein, Markus. 2012. *Approaching Human Performance*. PhD thesis, Springer.

Grebenstein, Markus, Alin Albu-Schäffer, Thomas Bahls, Maxime Chalon, Oliver Eiberger, Werner Friedl, Robin Gruber, Sami Haddadin, Ulrich Hagn, Robert Haslinger, et al. 2011. The DLR Hand Arm System. In *Proceedings of the IEEE International Conference on Robotics and Automation (ICRA)*, pp. 3175–3182.

Giordano, Paolo Robuffo, Matthias Fuchs, Alin Albu-Schäffer, and Gerd Hirzinger. 2009. On the Kinematic Modeling and Control of a Mobile Platform Equipped with Steering Wheels and Movable Legs. In *Proceedings of the IEEE International Conference on Robotics and Automation (ICRA)*, pp. 4080–4087.

Hogan, Neville. 1985. Impedance Control: An Approach to Manipulation: Part I-Theory, Part II–Implementation, Part III–Applications. *Journal of Dynamic Systems, Measurement, and Control* 107: 1–24.

Hogan, Neville. 1987. Stable Execution of Contact Tasks Using Impedance Control. *Proceedings of the IEEE International Conference on Robotics and Automation* 4: 1047–1054.

Hirzinger, Gerd, Norbert Sporer, Alin Albu-Schäffer, M Hähnle, Rainer Krenn, A Pascucci, and Manfred Schedl. 2002. DLR's Torque-Controlled Light Weight Robot III—Are we Reaching the Technological Limits Now? In *Proceedings of the IEEE International Conference on Robotics and Automation (ICRA)*, vol. 2, pp. 1710–1716. IEEE.

Hazara, Murtaza and Ville Kyrki. 2016. Reinforcement Learning for Improving Imitated In-contact Skills. In *Proceedings of the IEEE-RAS 16th International Conference on Humanoid Robots (ICHR)*, pp. 194–201. IEEE.

Khatib, Oussama. 1987. A Unified Approach for Motion and Force Control of Robot Manipulators: The Operational Space Formulation. *IEEE Journal on Robotics and Automation* 3 (1): 43–53.

Khatib, Oussama. 1986. Real-time Obstacle Avoidance for Manipulators and Mobile Robots. *International Journal of Robotics Research* 5 (1): 90–98.

Leidner, Daniel, Alexander Dietrich, Florian Schmidt, Christoph Borst, and Alin Albu-Schäffer. 2014b. Object-centered Hybrid Reasoning for Whole-body Mobile Manipulation. In *Proceedings of the IEEE International Conference on Robotics and Automation (ICRA)*, pp. 1828–1835.

Leidner, Daniel, Alexander Dietrich, Michael Beetz, and Alin Albu-Schäffer. 2016b. Knowledge-Enabled Parameterization of Whole-body Control Strategies for Compliant Service Robots. *Autonomous Robots (AURO): Special Issue on Whole-Body Control of Contacts and Dynamics for Humanoid Robots* 40 (3): 519–536.

Moro, Federico L, Michael Gienger, Ambarish Goswami, Nikos G Tsagarakis, and Darwin G Caldwell. 2013. An Attractor-based Whole-body Motion Control (WBMC) System for Humanoid Robots. In *Proceedings of the IEEE-RAS International Conference on Humanoid Robots (ICHR)*, pp. 42–49.

Nakamura, Yoshihiko, Hideo Hanafusa, and Tsuneo Yoshikawa. 1987. Task-Priority Based Redundancy Control of Robot Manipulators. *International Journal of Robotics Research* 6 (2): 3–15.

Ott, Christian, Alexander Dietrich, Daniel Leidner, Alexander Werner, Johannes Englsberger, Bernd Henze, Sebastian Wolf, Maxime Chalon, Werner Friedl, Alexander Beyer, et al. 2015. From Torque-ontrolled to Intrinsically Compliant Humanoid Robots. *Mechanical Engineering* 3 (2): 7–11.

Ott, Christian. 2008. *Cartesian Impedance Control of Redundant and Flexible-Joint Robots*, vol. 49. Springer Tracts in Advanced Robotics Berlin Heidelberg: Springer Publishing Company.

Ott, Christian, Alin Albu-Schäffer, Andreas Kugi, and Gerd Hirzinger. 2008. On the Passivity-based Impedance Control of Flexible Joint Robots. *IEEE Transactions on Robotics* 24 (2): 416–429.

Ott, Christian, Alin Albu-Schäffer, Andreas Kugi, Stefano Stramigioli, and Gerd Hirzinger. 2004. A Passivity Based Cartesian Impedance Controller for Flexible Joint Robots—Part I: Torque Feedback and Gravity Compensation. In *Proceedings of the IEEE International Conference on Robotics and Automation (ICRA)*, pp. 2659–2665.

Ortenzi, Valerio, Maxime Adjigble, Kuo Jeffrey, Rustam Stolkin, and Michael Mistry. 2014. An Experimental Study of Robot Control During Environmental Contacts based on Projected Operational Space Dynamics. In *IEEE-RAS International Conference on Humanoid Robots (ICHR)*, pp. 407–412.

Siciliano, Bruno and Jean-Jacques Slotine. 1991. A General Framework for Managing Multiple Tasks in Highly Redundant Robotic Systems. In *Proceedings of the International Conference on Advanced Robotics (ICAR)*, pp. 1211–1216.

Sentis, Luis, and Oussama Khatib. 2005. Synthesis of Whole-body Behaviors Through Hierarchical Control of Behavioral Primitives. *International Journal of Humanoid Robotics* 2 (4): 505–518.

Sadeghian, Hamid, Luigi Villani, Mehdi Keshmiri, and Bruno Siciliano. 2014. Task-Space Control of Robot Manipulators with Null-Space Compliance. *IEEE Transactions on Robotics* 30 (2): 493–506.

Schindlbeck, Christopher and Sami Haddadin. 2015. Unified Passivity-Based Cartesian Force/Impedance Control for Rigid and Flexible Joint Robots Via Task-Energy Tanks. In *Proceedings of the IEEE International Conference on Robotics and Automation (ICRA)*, pp. 440–447.

Silvério, Joao, Leonel Rozo, Sylvain Calinon, and Darwin G Caldwell. 2015. Learning Bimanual End-Effector Poses from Demonstrations Using Task-Parameterized Dynamical Systems. In *IEEE/RSJ International Conference on Intelligent Robots and Systems (IROS)*, pp. 464–470. IEEE.

Tan, Jindong, Ning Xi, and Yuechao Wang. 2003. Integrated Task Planning and Control for Mobile Manipulators. *The International Journal of Robotics Research* 22 (5): 337–354.

Vanthienen, Dominick, Steven Robyns, Erwin Aertbeliën, and Joris De Schutter. 2013. Force-Sensorless Robot Force Control Within the Instantaneous Task Specification and Estimation (iTaSC) Framework. In *Benelux Meeting on Systems and Control*.

Wolf, Sebastian, Oliver Eiberger, and Gerd Hirzinger. 2011. The DLR FSJ: Energy Based Design of a Variable Stiffness Joint. In *Proceedings of the IEEE International Conference on Robotics and Automation (ICRA)*, pp. 5082–5089. IEEE.

Wimböck, Thomas. 2013. *Controllers for Compliant Two-Handed Dexterous Manipulation*. PhD thesis, Vienna University of Technology.

Yoshikawa, Tsuneo. 1990. *Foundations of Robotics: Analysis and Control*. Cambridge: The MIT Press.

Yamamoto, Yoshio and Xiaoping Yun. 1992. Coordinating Locomotion and Manipulation of a Mobile Manipulator. In *Proceedings of the IEEE Conference on Decision and Control (CDC)*, pp. 2643–2648.

Chapter 7
Semantic Interpretation of Haptic Feedback

The previous chapters developed the representations, the planning methods, and a suitable approach to execute compliant manipulate tasks at the example of wiping chores. However, up to this point, the robot is still unaware of the actually achieved task performance. To that end, this chapter investigates the last remaining element of the Intelligent Physical Compliance concept, i. e. the interpretation of the executed actions and the subsequent estimation of the task performance. This closes the cognitive control loop (Norman and Shallice 1980) as it is depicted in Fig. 7.1. The interpretation of effects is usually considered a machine vision problem. However, visual data is often unreliable for wiping actions for two main reasons. First, the medium in wiping tasks may be invisible due to lighting conditions or the properties of the medium such as small dust and dirt particles, or transparent liquids. Second, the simulations performed in Chap. 5, as well as the experiments conducted in Chap. 6 indicate that the robot itself is often occluding the manipulated areas which makes visual perception of the task outcome only available after the task execution. For this reasons, this chapter presents a vision independent method to interpret the effect of wiping actions based on haptic feedback.

The remainder of this chapter is structured in two main sections. First, the effect inference strategy to estimate the task performance of real world wiping motions based on haptic feedback is introduced in Sect. 7.1. Second, a concept for semantic annotation of episodic memories and the integration with the openEASE framework is described in Sect. 7.2. The findings in this chapter are based on the research published in the conference paper (Leidner and Beetz 2016).

D. S. Leidner, *Cognitive Reasoning for Compliant Robot Manipulation*, Springer Tracts in Advanced Robotics 127,
https://doi.org/10.1007/978-3-030-04858-7_7

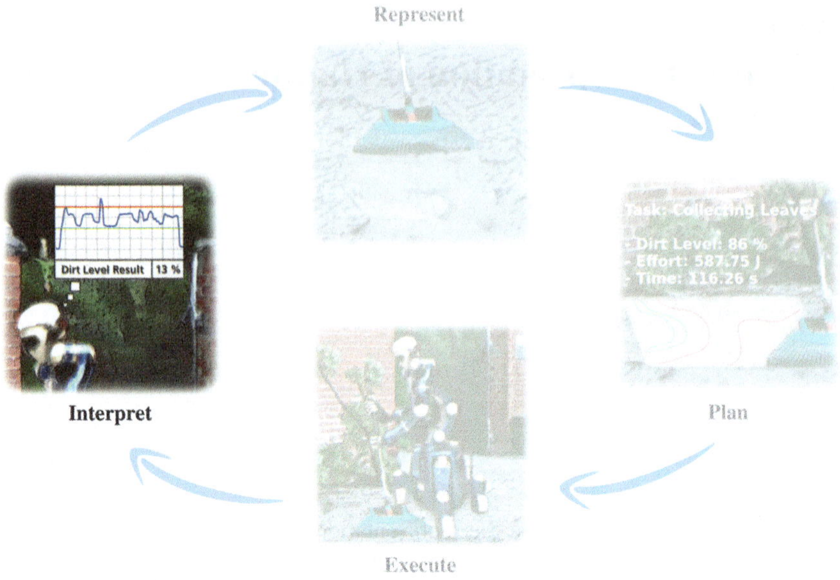

Fig. 7.1 Interpreting Intelligent Physical Compliance

7.1 Effect Inference Based on Haptic Perception

Service robots are envisaged as universal assistants in everyday environments. As such, they will have to cope with a wide variety of daily household chores, including cooking, organization and cleaning. Within the context of robotic manipulation, but especially for cleaning tasks, it is important to monitor the performance of the executed actions. Typically, humans evaluate the outcome of their actions based on visual perception. For example, collecting shattered shards of a broken mug with a broom will eventually result in an accumulated pile of shards visually noticeable at the dedicated goal region. Another example is the absorption of larger dirt particles with a vacuum cleaner, which results in a clean surface if all particles are removed. Based on these assumptions, there have been some efforts on effect-oriented robotic cleaning in the past (Hess et al. 2011; Martínez et al. 2015; Do et al. 2014). However, visual feedback is an unreliable source of information. Even though visual perception is commonly considered to be the prime sense in human manipulation, visual feedback is unavailable for many tasks including vacuuming the floor, dusting surfaces, and window wiping, as small dirt particles, dust, and streaks of water are hardly perceivable in camera images or depth images, especially on transparent and reflecting surfaces.

To overcome this issue, humans do not solely rely on perceptual feedback, but also maintain knowledge of their manipulation actions and the resulting effects in form of abstract process models according to research on neurobiology (Kawato 1999).

For example, in order to clean a sideboard with a feather duster, a human would try to cover the entire surface of the target object by wiping along it with the tool. This behavior is based on the knowledge that dust particles are electrostatically absorbed by the feather duster and the assumption that the dust is equally distributed on the planar surface. Consequently, humans are able to infer that the desired effect (i. e. having the side board cleaned from dust) is successfully accomplished after the tool has been in physical contact with the whole target area.

Nevertheless, the removal of the dust is not visible and the effect is not directly perceivable. The only reliable feedback left is *haptic perception* (Gibson 1966), i. e. the active exploration of the environment using proprioception and force information as it occurs from contact during wiping motions. The sense of touch is an essential factor for the task reasoning and effect inference of humans in the absence of vision (Gibson 1962). Haptic feedback provides humans not only with the information that contact was successfully established, but also provides the basis to rate the task performance and even detect performance errors based on the comparison of the desired contact forces and the actual sensed force (Flanagan et al. 2006). In case of irregular contact situations (i. e. introduced by friction or uneven areas) humans may decide to replan additional wiping motions to revisit the affected regions and improve the cleaning result.

To this end, the combination of cognitive capabilities and haptic perception enables humans to qualitatively reason about the effects of their motions and solve even complex cleaning tasks despite poor visual perception. This section proposes to utilize the torque sensing capabilities of compliant light-weight robots (Hirzinger et al. 2002) to infer contact situations of compliant wiping motions and measure the task performance based on a qualitative effect model. This way, a robot is not just able to estimate the performance of its actions qualitatively, but it is also able to question its own actions and schedule repair motions in case of the detection of insufficient contact.

The steps to close the loop from semantic reasoning to low-level control and vise versa is outlined in Fig. 7.2. The entire procedure is based on the particle distribution model introduced in Sect. 4.3. In a nutshell, the robot executes wiping motions according to the planing methods introduced in Chap. 5 (top). As the robot executes the planned wiping motions by means of a compliant whole-body impedance, it is continuously recording telemetry data. This includes the controller force $f$cmd as it is calculated from joint torque measurements (6.10), as well as measured Cartesian end-effector positions $\boldsymbol{x}_{\text{act}}$. With this information the robot is able to reproduce the executed motions in simulation and estimate the contact between the tool and the particle distribution w. r. t. real world sensor readings and compute the real task performance respectively (bottom). The measurements that match the desired force profile are considered in contact with the target surface (green bars). Whenever contact is detected, that is not aligned with the surface, it is considered a disturbance (red bars). Based on this information the effect to the particle distribution is simulated in order to evaluate the real world task outcome. The resulting particle distribution can be used to reinitialize the planning methods and compensate for execution errors. The individual reasoning steps are detailed in the following subsections.

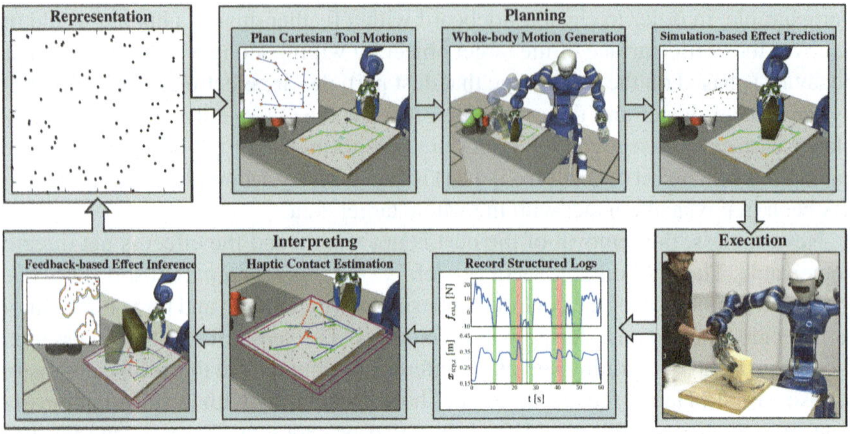

Fig. 7.2 Coupling between the planning approach and the interpretation of wiping motions. The particle distribution model serves as the basis for the earlier introduced motion planning algorithm as well as the haptic effect inference method discussed in this chapter. The methods complement each other such that the output of the effect inference can be utilized to plan recovery motions in case of estimated performance errors introduced by external disturbances, e. g. collisions with humans as visualized

7.1.1 Contact Estimation

Provided that a robot planned its wiping motions based on the reasoning methods introduced in Chap. 5, and executes the motions by means of the compliant whole-body impedance control framework introduced in Chap. 6, it is possible to estimate contact with the environment based on *haptic feedback*. To do so, the actual measured end-effector position $\boldsymbol{x}_{\text{act}}$ and the controller forces at the end-effector $\boldsymbol{f}_{\text{cmd}}$ are recorded at each timestamp i (cf. 6.10). To give an illustration, an example trajectory of the computed TCP position is visualized as black dotted line in Fig. 7.3. The blue lines on the target surface indicate the desired wiping motions of the sponge TCP in contact with the chopping board. The visualized overall motion is the outcome of a collect action planned based on the RRT coverage strategy (see Sect. 5.3.1), where an arrow-shaped path is formed pointing toward the goal node n_{goal}.

Each TCP position is related to the corresponding contact force measurement provided by the recorded log. This allows to infer the segments of motion that are most likely in contact with the environment. In fact, only the forces normal to the target surface $\boldsymbol{f}_{\text{ext, n}}$ are of interest. The normalized contact force

$$f'_{\text{n}, i} = \frac{\min(f_{\text{ext, n, max}}, \; f_{\text{ext, n}, i})}{f_{\text{ext, n, min}}}, \tag{7.1}$$

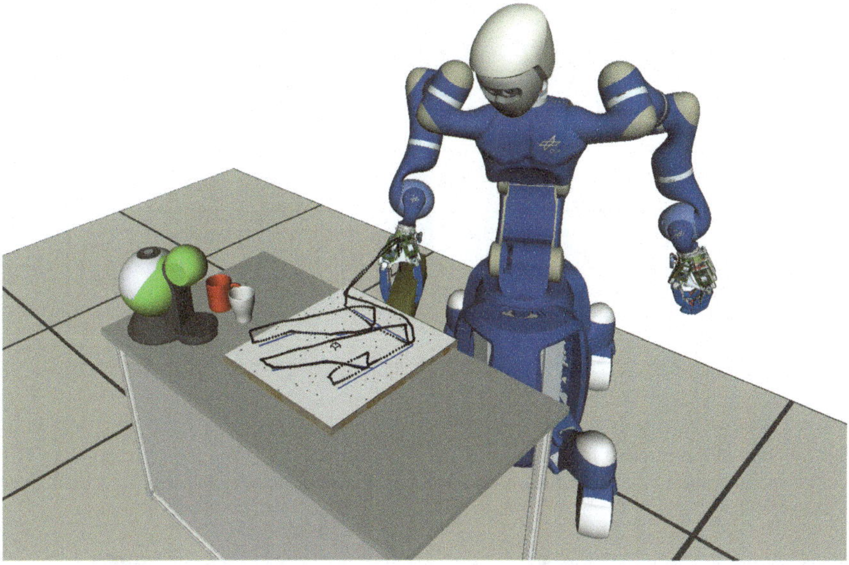

Fig. 7.3 Visualization of a recorded collect motion executed by Rollin' Justin. The recorded Cartesian wiping motion of the robotic manipulator, i. e. the TCP of the sponge it is holding respectively are plotted as black dotted line. The desired wiping motion in contact is shown as blue lines

for each force sample $f_{\text{ext, n}, i}$ are analyzed. In the example at hand $f_{\text{ext, n, max}} = 0\,\text{N}$ and $f_{\text{ext, n, min}} = -10\,\text{N}$. Accordingly, positive force values arising from gravitational effects (i. e. lifting the robots own weight) are explicitly neglected.

Only the TCP positions that show a high confidence for contact with the chopping board are considered in the first place. In other words, only measurements that show a normalized force value of [0.9, 1.0] are investigated, such that

$$\boldsymbol{x}'_{\text{tcp}, i} = \left\{\boldsymbol{x}_{\text{tcp}, i} \,|\, 0.9 \leq f'_{\text{n}, i} \leq 1.0\right\}, \tag{7.2}$$

i. e. the measurements that show only 10% deviation from the desired contact force as it is emphasized in Fig. 7.4. The corresponding motion is visualized by green dots in the simulation presented on the left of Fig. 7.5. Most of the measurements close to the target surface match this constraint and resemble the desired wiping trajectory. This is already a quite accurate estimate of the contact motion, yet, some segments were omitted due to lower contact forces introduced by friction and stick-slip effects. However, these segments cannot be ignored as they participate to the overall wiping effect.

These left out yet still contact-rich segments can be incorporated by applying the *Random Sample Consensus (RanSaC)* algorithm (Fischler and Bolles 1981). The RanSaC algorithm is often used in computer vision to fit a plane onto surface elements perceived in visual data, e. g. point clouds computed based on depth camera images. Here it is applied on the Cartesian TCP positions with high contact confidence $\boldsymbol{x}'_{\text{tcp}, i}$

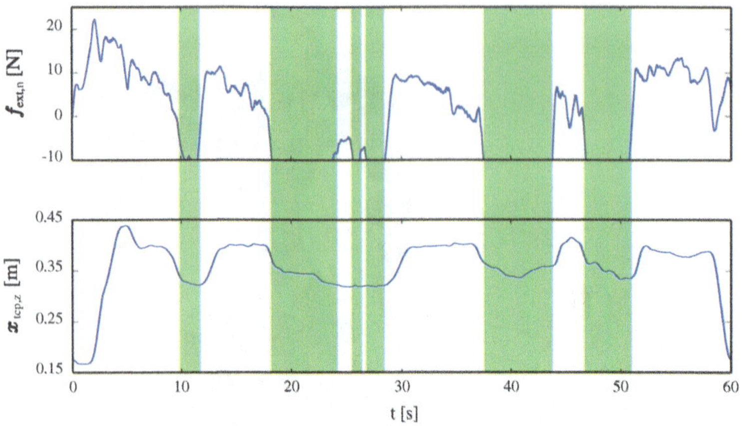

Fig. 7.4 Plot of the external force normal to the target surface $f_{\text{ext, n}}$ and the corresponding TCP position in z-direction $x_{\text{tcp, z}}$. The measurements with high contact confidence $x'_{\text{tcp},i}$ are highlighted by the green bars

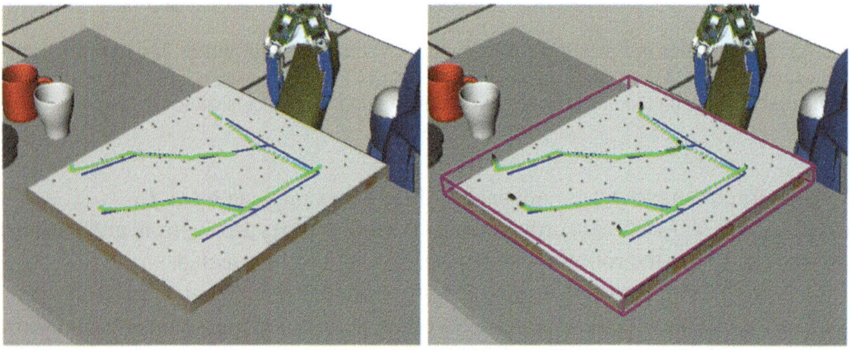

Fig. 7.5 Left: Closeup view of all measurements with high contact confidence $x'_{\text{tcp},i}$ (green dots), which are the basis for the target surface estimation. Right: The estimated target surface visualized as purple box. All positions of the TCP within this box $x_{\text{plane}, i}$ are considered in contact. The brighter the green color, the higher the normalized contact force $f'_{\text{n}, i}$

to fit a plane onto the target surface, such that

$$x_{\text{plane}, i} = \text{RanSaC}\left(\sum_{i=0}^{N} x'_{\text{tcp}, i}, \epsilon\right), \tag{7.3}$$

where $x_{\text{plane}, i}$ constitute the measurements within the estimated target plane visualized as purple box in Fig. 7.5. All measurements $x_{\text{plane}, i}$ that fit within the inlier threshold ϵ along the estimated surface orientation are considered in contact with the surface. The corresponding measurements are visualized as green dots of different brightness. High normalized contact forces $f'_{n,i}$ are represented by bright green colors. Darker green colors (eventually fading to black) represent lower normalized contact forces $f'_{n,i}$. The RanSaC based approach incorporates all measurements that are potentially in contact with the target surface for contact modeling, and not solely the data points with high contact confidence. This way of haptic perception allows the robot to estimate the target surface despite poor lighting conditions as they may occur on transparent glass panes or reflecting objects, for example.

7.1.2 Effect Inference

The model for the simulation-based effect prediction in Sect. 5.3.5 does only consider the volumetric dimension of the tool in relation to the position of the particles. There are no forces involved to simulate the wiping effect. However, in real world applications lower contact forces may result in poor contact situations. Therefore, this section shows how the normalized contact force $f'_{n,i}$ can be incorporated to model the effect of real world wiping actions. In particular, the enhanced effect model is based on

- the position of the TCP of the tool in contact $\boldsymbol{x}_{\text{plane},i}$,
- the particle distribution $\mathcal{P}$ w. r. t. the tool CAD data,
- and the applied controller force f_{cmd}, respectively the external force f_{ext}.

By re-executing the recorded motions in simulation the robot is able to infer the real world effect of the prior executed wiping motions under consideration of the logged contact forces. It is argued that contact forces close to the desired force have higher probability to produce the desired wiping effect to the medium. However, the actual effect depends on the properties of the tool-medium-surface tuple. Among others, the exact tool geometry, the tool elasticity, the surface friction, the surface inclination, the medium friction, and the medium size and geometry influence the outcome of wiping actions. Most of these parameters are hardly assessable. Therefore, the *log-likelihood function* $\log(\mathcal{L}(\theta|x))$ is utilized to model the effect on a qualitative basis. At every timestamp, the log-likelihood is computed for each particle in contact with the tool.

- If $f'_{n,i} \geq ||\log(\mathcal{L}(\theta|x_i))||$, the contact behavior is simulated as described in Sect. 4.3 (i.e. push the affected particles in case of collect and skim actions and delete the particles for the absorb action).
- If $f'_{n,i} < ||\log(\mathcal{L}(\theta|x_i))||$, the simulation step is skipped without applying the effect and the algorithm proceeds with the next measurement.

The *probability density function* $p(x)$ for the log-likelihood function is plotted in Fig. 7.6. It is defined as

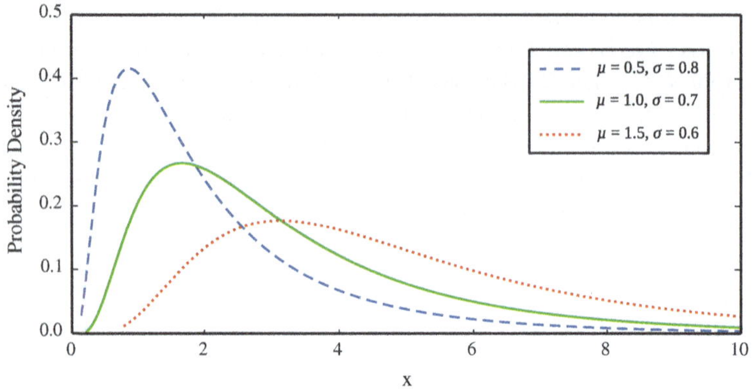

Fig. 7.6 Example plots of the probability density function for the log-likelihood function for different contact situations

$$p(x) = \frac{1}{x\,\sigma\,\sqrt{2\,\pi}}\,e^{-\dfrac{(\log(x)\;-\;\mu)^2}{2\,\sigma^2}}, \tag{7.4}$$

where μ is the mean and σ is the standard deviation of the logarithm. These variables can be altered to simulate contact models for different tool-medium-surface combinations with varying properties. This approach avoids a fixed force threshold by exploiting the variance of the likelihood function. While a fixed force threshold may be sufficient to distinguish contact from no contact, it will often result in false positives in borderline situations. Utilizing a steeply parameterized log-likelihood function, even path segments with a lower force measurement have a chance to produce the desired outcome (e. g. to simulate a window wiper skimming water from a window, blue, dashed line). Vice versa, even high contact forces may result in no effect for more flat log-likelihood functions (e. g. to simulate the bristles of a broom collecting fine sand, red, dotted line).

A collect action with a sponge is shown in Fig. 7.7, where small chippings of glass are distributed on a chopping board. This task is based on the simulations conducted in Sect. 5.3.5. The real outcome is shown at the top. The resulting effect estimation over time is visualized below. Please note that the initial particle distribution is not visually perceived (which is of course always an option for visible particles or liquids). Similarly to the simulations conducted in Sect. 5.3.5, a unified particle distribution is assumed to emulate the absence of visual feedback. The minor contact loss observed in Figs. 7.4 and 7.5 does not affect the overall particle distribution estimation. Accordingly, all particles are collected on the lower right corner of the target surface.

In the following it is shown how the log-likelihood function can be adapted to simulate the behavior for a different tool-medium-surface constellation. Similarly to the fictional example of collecting leaves with a rake (cf. Fig. 1.1) and the conducted experiment of collecting shards with a broom (cf. Fig. 6.10), the second scenario is

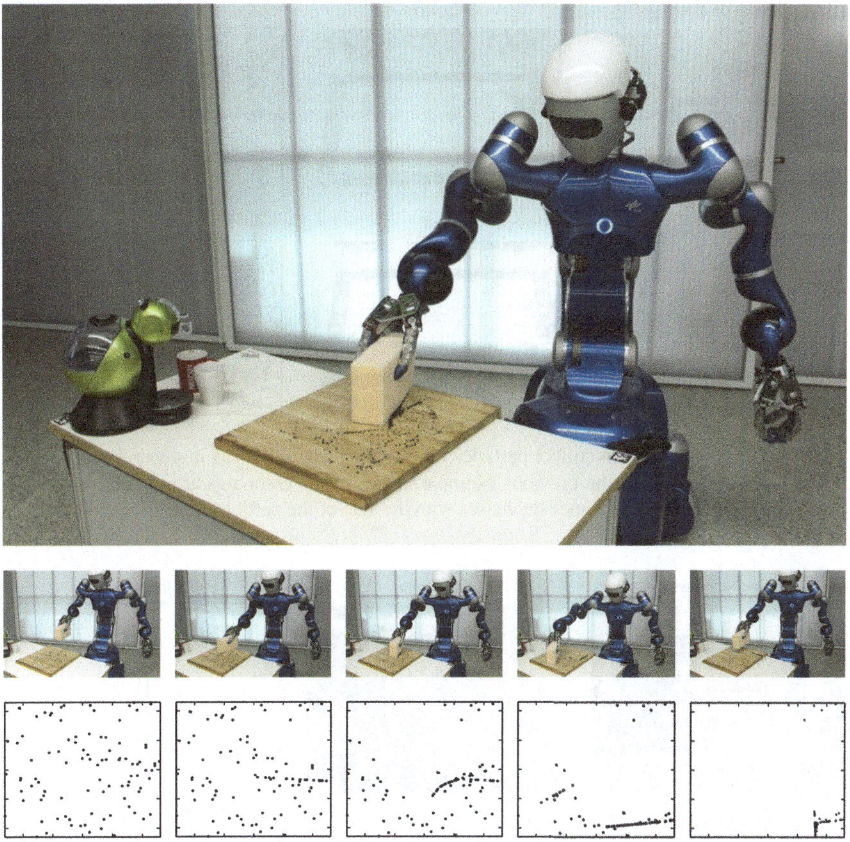

Fig. 7.7 The real collect action with a sponge executed by Rollin' Justin (middle) compared to the estimated outcome (bottom). The particle distribution in the bottom row is rotated clockwise, where the upper left corner corresponds to the left corner of the chopping board in the top row. Black chippings of glass are placed to visualize the task performance and assess the results of the effect inference method. They are not visually perceived by the robot. Instead a uniform initial distribution is assumed

based on the bi-manual example of collecting dirt particles on the floor using a broom (see Fig. 7.8). This experiment is executed twice with particles of different kind. The robot motion as well as the controller parameterization is identical in both trials. First, the chippings (2–6 mm) introduced in the previous example are distributed on a sheet of paper on the floor. As the broom swipes over the surface most of the particles are effected. Few chippings remain as the bristles of the broom are of irregular nature. Second, small grains of sand (0.2–2 mm) are distributed. These particles are too small to be efficiently collected by the broom as the bristles bend. This effect can be modeled with the particle representation by utilizing a flatter log-likelihood function.

The third scenario investigates the effect of different tools to a certain medium. In particular, the robot is commanded to skim detergent from a glass panel utilizing

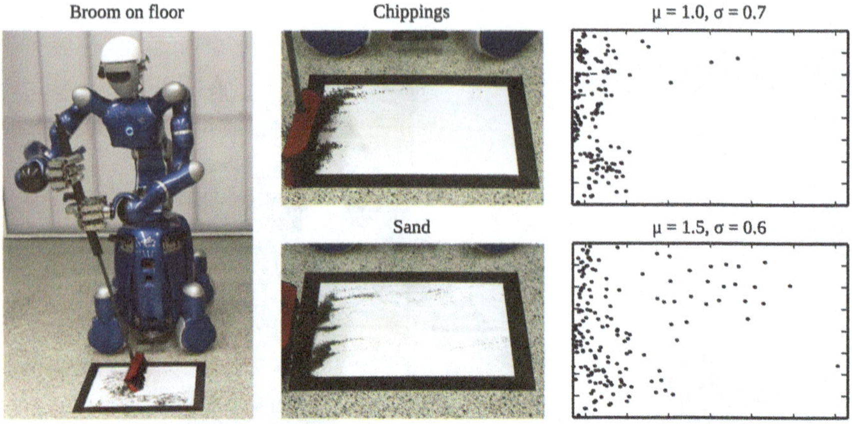

Fig. 7.8 A broom is used to collect particles on the floor. The broom is in general less accurate than the sponge utilized in the previous example. Some of the chippings are not effected by the broom (top). The task performance decreases with the size of the particles as it is observed for the fine grained sand (bottom)

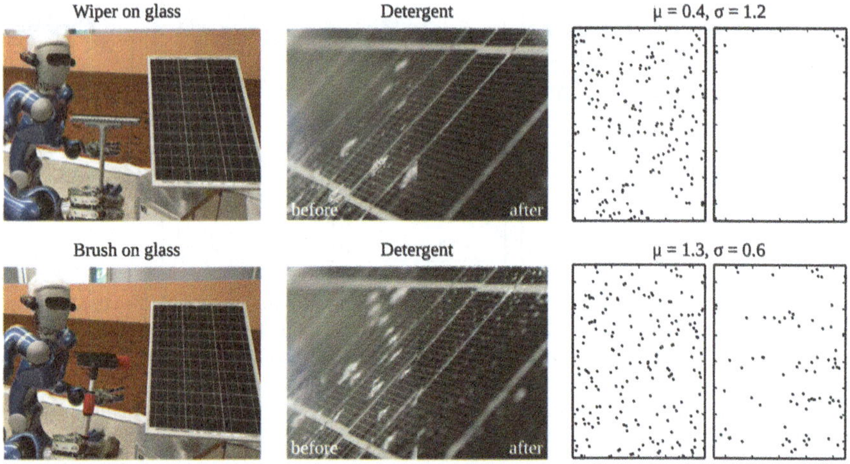

Fig. 7.9 Detergent is applied to a glass panel. The robot executes a skim action with a window wiper and a brush in order to remove the liquid. The detergent on the glass panel is shown before (left) and after (right). The model parameters μ and σ are designed to match the effect of the two tools on the medium

the rubber blade of a window wiper in comparison to the bristles of a brush (see Fig. 7.9). Since, the brush is not designed to manipulate liquids, the window wiper outperforms the brush in this task naturally. Macro recordings before and after the wiping action have been captured to better assess the effect. One can see that the window wiper removes most of the liquid from the glass panel surface, whereas a layer of detergent remains utilizing the brush. Similarly to the other experiments, this circumstance can be modeled by adapting the parameters of the log-likelihood function. Even though the detergent is hardly visible, the particle model allows to estimate the effect qualitatively.

7.1.3 Failure Detection and Recovery

As already described, the wiping motions in this work are executed by means of a whole-body impedance controller. The applied contact force is thereby a side-effect from the controller force that results from the commanded torque (6.10). The controller force is thereby saturated w. r. t. the requirements of the wiping action. This approach is proven to be suitable for wiping actions (Hogan 1987) and allow for safe human-robot interaction likewise (Albu-Schäffer et al. 2007a). However, as the controller is not designed to adapt the Cartesian position w. r. t. the force measurements (unlike a hybrid position-force controller as described in (Denei et al. 2015), it is prone to errors arising from imprecise localization (see Fig. 7.10 left) and external perturbation (see Fig. 7.10 right). Nevertheless, the proposed inference method is able to detect these execution errors and adapt accordingly.

The first investigated failure scenario is the loss of contact due to imprecise localization. In particular, the table in front of the robot is tilted to emulate a rotational localization error. The commanded wiping motions are equal to the successful task execution. A plot of the contact force and the TCP position is provided in Fig. 7.11. The corresponding TCP motion and the plane estimation is visualized in Fig. 7.12. The plot already hints at the fact that only few confident contact situations occurred during the task execution. This gets more obvious in the simulation. While the left segments of the arrow-shaped path roughly match the perceived table height, the path segments on the right are rendered too high and show low contact forces (dark green and black dots). The purple box is visualizing the plane estimation is shifted

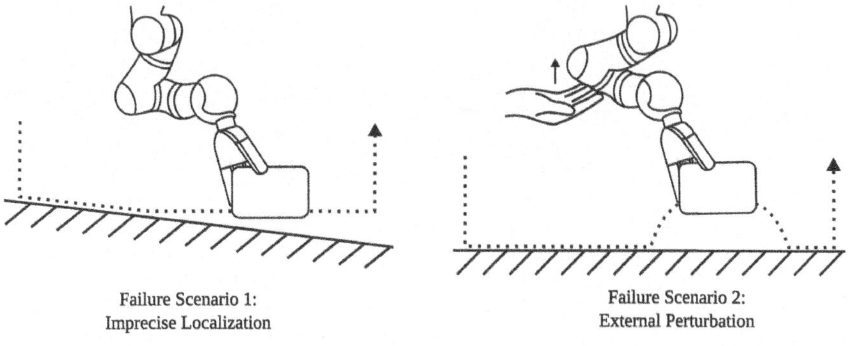

Fig. 7.10 Illustration of the two investigated failure scenarios. First, contact loss due to imprecise localization (left). Second, contact loss due to external human perturbation (right)

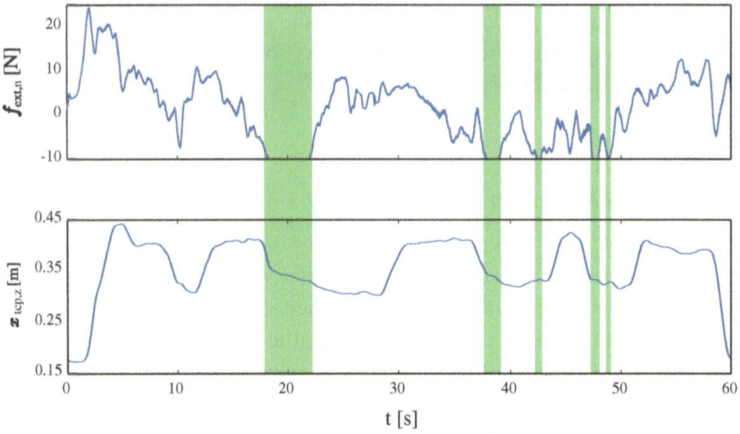

Fig. 7.11 A plot of the contact force $f_{ext,\,n}$ and the TCP position in z-direction $x_{tcp,\,z}$ in a failure situation. While the position stays on certain level, it is notable that the contact force is of irregular nature. The contact loss is evident in the reduced number of confident contact situations $x'_{tcp,\,i}$ (green bars)

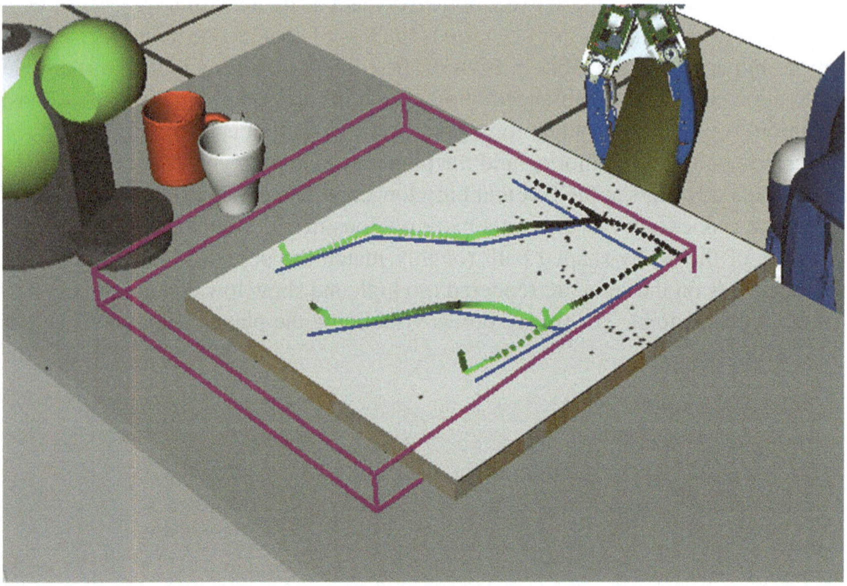

Fig. 7.12 The recorded Cartesian wiping motion in a failure situation arising from a tilted table. The green and black path visualizes the actual measured TCP trajectory. The commanded path is visualized as blue lines. The few confident contact measurements are sufficient to estimate the target surface with the RanSaC algorithm. However, the limited contact force results in less effective motions, such that the particles close to the table edge remain unmoved in the simulation

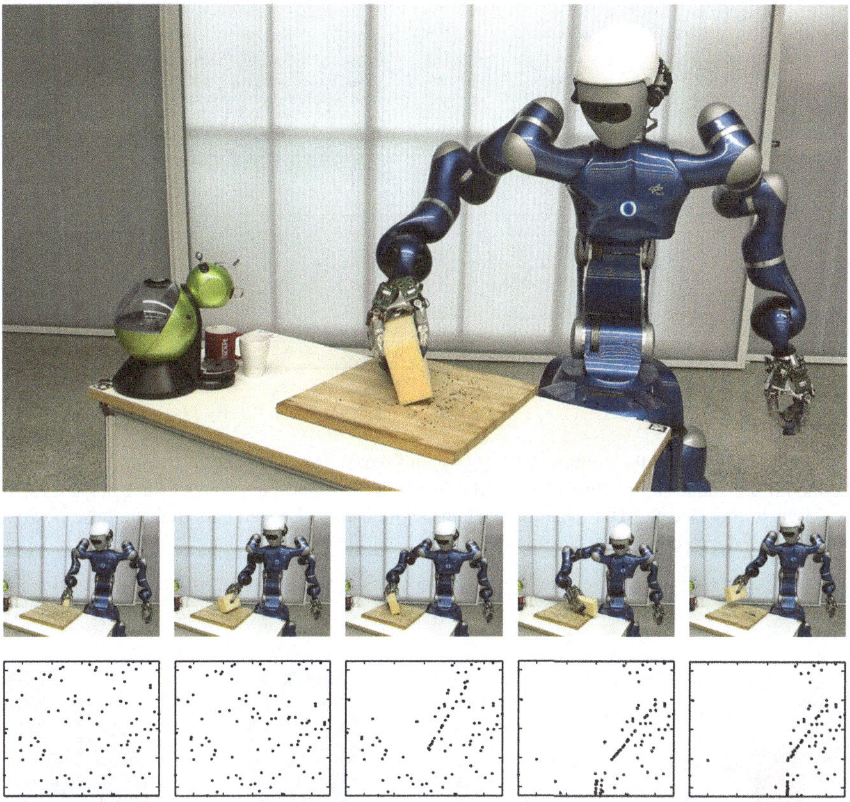

Fig. 7.13 Failure Scenario I: The tilted table simulates an incorrect localization leading to partially bad contact situations shown in the top row. The estimated particle distribution is shown in the lower row. Similarly to the first example the particle behavior is simulated w. r. t. the real world telemetry data recorded during the task execution

to the right accordingly. Figure 7.13 shows the task execution in five snapshots. The contact loss is captured in the fourth frame. A simple comparison of the volumetric model of the sponge and the particle distribution would not be sufficient to estimate the contact situation as the sponge is still very close to the target surface. By referencing the measured tool motion to the force at the end-effector the robot is able to detect the contact loss and model the effect as visualized in the lower row. The final estimation of the particle model is shown to be very similar to the real breadcrumb distribution on the chopping board.

The second failure scenario showcases a deliberate human intervention. The robot is pushed up, such that the sponge looses contact with the chopping board. As the maximum controller force is saturated to satisfy the contact behavior, the robot will not counteract the human and the controller force measurement will not differ from the nominal case. However, the position of the end-effector does not match

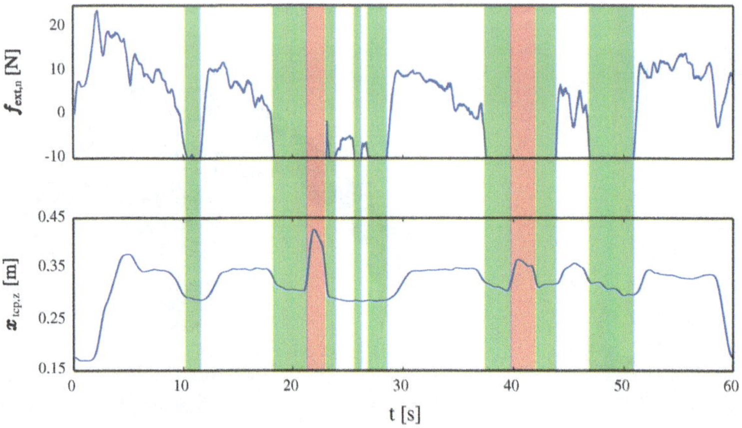

Fig. 7.14 The plot for the contact force $f_{\text{ext, n}}$ and the TCP position in z-direction $x_{\text{tcp, z}}$ in the second failure situation. In comparison to the first failure scenario the contact force stays leveled. However, in this case the TCP position deviates significantly. The red bars indicate false positive contacts introduced due to this perturbation

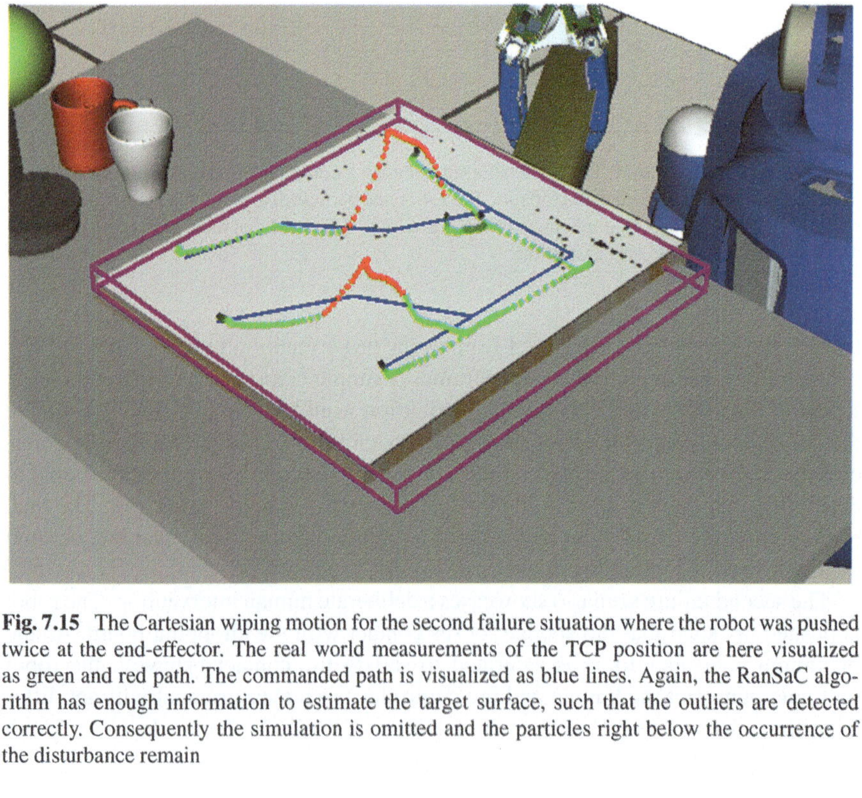

Fig. 7.15 The Cartesian wiping motion for the second failure situation where the robot was pushed twice at the end-effector. The real world measurements of the TCP position are here visualized as green and red path. The commanded path is visualized as blue lines. Again, the RanSaC algorithm has enough information to estimate the target surface, such that the outliers are detected correctly. Consequently the simulation is omitted and the particles right below the occurrence of the disturbance remain

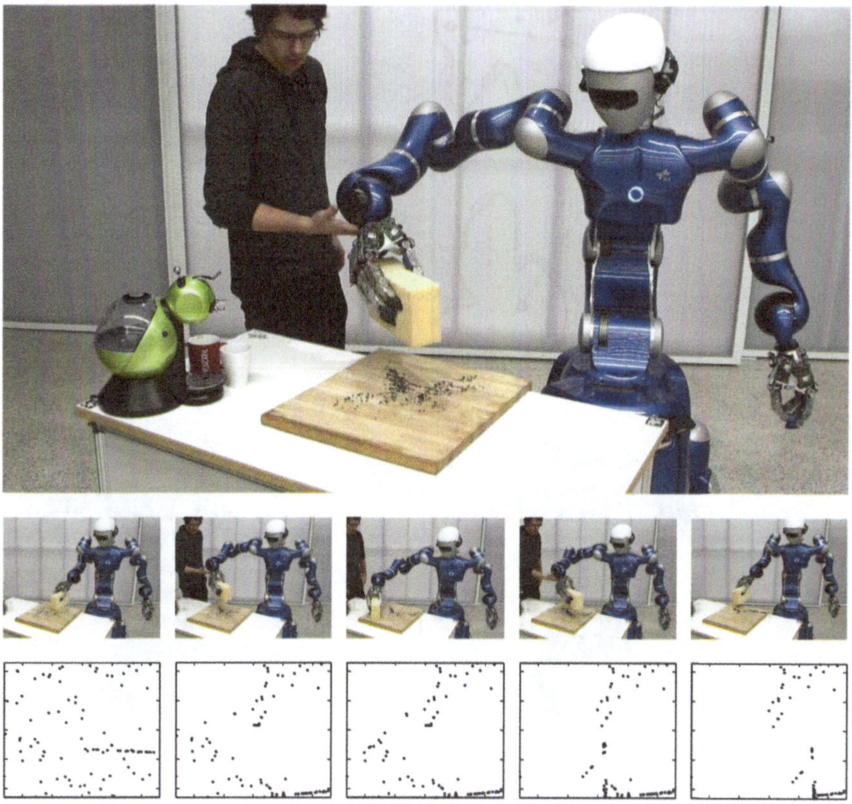

Fig. 7.16 Failure Scenario II: The robot is pushed twice at the end-effector during task execution as it is shown in the top row. The estimated particle distribution visualized in the bottom row reflects this disturbance

the desired position. Thanks to the RanSaC algorithm used to estimate the target plane, the approach is able to detect these outliers and ignore them during the effect estimation procedure. Similarly to the first failure scenario the robot is eventually able to correctly infer where the wiping motions have been efficient, and where the desired effect was not carried out. The false positive measurements are visualized as red bars in Fig. 7.14 and the matching red path segments in Fig. 7.15. In the plot one can see that the contact force stays almost constant during the disturbance while the position is significantly offset. The corresponding interaction is visualized in frame two and four in Fig. 7.16. This interaction right before the two main intersections of the wiping motions is a significant impairment for the overall task performance. Some particles are estimated to remain on the chopping board similarly as observed for the real execution.

As argued in the introduction of this chapter, humans are capable of detecting execution errors based on haptic perception, update their internal task models accordingly, and use the new information to recover from the failure situation. The represen-

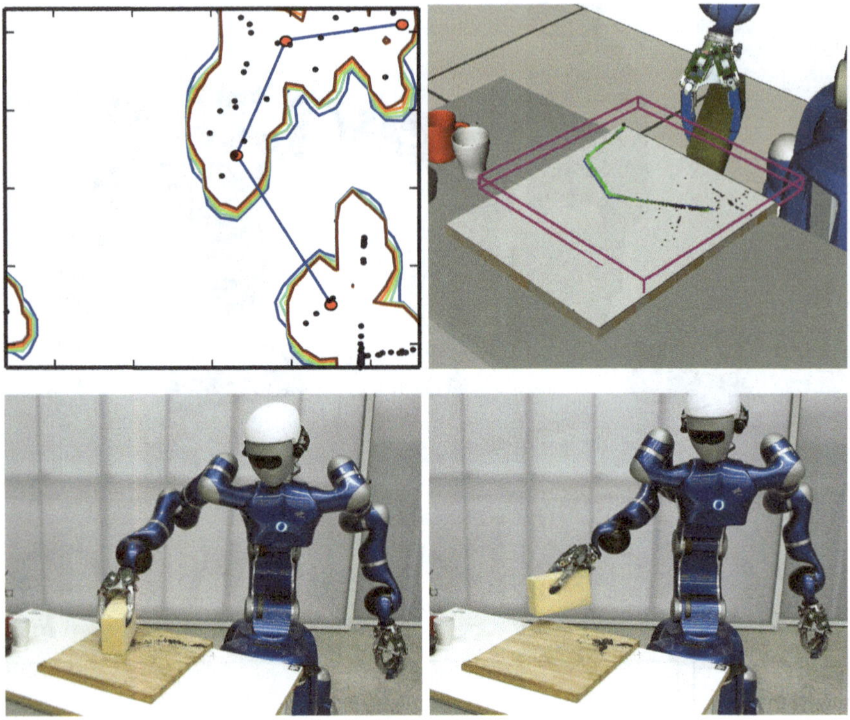

Fig. 7.17 Top left: KDE for the remaining particle distribution after the second failure scenario. Top right: The recorded recovery motion and the final particle distribution. Bottom: The eventually successful real world execution

tation of wiping motions and their effects enables a robot to close this cognitive loop in a similar way for wiping tasks. In particular, utilizing the output of the effect inference method to replan additional wiping motions allows the robot to recover from possible failure situations introduced due to bad contact situations. Consequently, the robot is able to plan the recovery motion directly in the effect-space. As the effect inference is mostly matching the task performance of the real world execution, there is no visual feedback required. This makes the proposed approach also applicable to wiping tasks involving transparent liquids or small dirt particles, e. g. water or dust.

The recovery procedure for the second failure scenario is outlined in Fig. 7.17. The initial particle distribution to plan the recovery motion is based on the final estimation of the particle distribution after the robot was pushed twice at the right manipulator (cf. Fig. 7.16). To plan the recovery motion most effectively, the KDE strategy is used to distribute the nodes for the SDG w. r. t. the regions with high particle density (red dots in the top left) and replan the collect action accordingly. Similarly to the previous executions, the robot generates whole-body joint motions, executes them, and records structured logs during run-time. This enables the robot to re-evaluate the task performance w. r. t. the new measurements and infer the improvement qualitatively. In the example at hand, the robot increased the performance for the correctly accumulated particles from 80 to 97 % w. r. t. the performance metric for collect actions (4.3). Similarly, a recovery plan could be applied to the first failure scenario (after the robot is commanded to relocalize itself) or any other scenario with uncertain contact situations to improve arbitrary everyday wiping tasks.

7.1.4 Discussion

In conclusion, the applied computational model presents a suitable estimation of the real world effect. Even though the contact behavior poses an abstraction of the actual process (i. e. the contact between the tool-medium-surface tuple as well as the motion of the particles are strongly simplified), the resulting patterns in the particle model match the real world observations in general which allows for a qualitative assessment.

Although the results of the effect inference are quite satisfactory, the parameterization of the effect model introduces some limitations. For now, the tool dependent parameters for the log-likelihood function (i. e. μ and σ), as well as the maximum allowed contact force $f_{\text{ext, n, max}}$ have been defined empirically. In contrast, it was shown that task parameters in the context of wiping motions can be learned over time based on the visual observation of the effect (Do et al. 2014) or human demonstration (Hazara and Kyrki 2016). Given the right parameterization, the applied effect model roughly matches the real world outcome for the showcased scenarios. Even though the applied effect model is only a qualitative representation of the actual procedure, it can be used for a first quality estimation in real-time. A more accurate estimation of the physical behavior could be conducted by a physics simulation such as YadeDEM (Šmilauer et al. 2010), which is able to simulate the contact behavior between spherical or clustered particles and objects of arbitrary geometric shape in three-dimensional space.

7.2 Semantic Analysis with OpenEASE

A truly cognitive-enabled system should not only be able to reason about actions and effects numerically, but also on higher levels of abstraction, such that a human operator can query the system semantically, e. g. by means of natural language. Especially for fully automated processes in industrial settings this is extremely valuable as big data is generated over long time periods. An operator cannot monitor the process end-to-end nor search through endless data streams to investigate performance errors. Instead, the increasing complexity requires to query the system based on specific interest, such as quality control or safety aspects. Possible queries for industrial settings could include requests like *Visualize all motions with an estimated task performance of 80 % or lower.* or *Highlight all motion segments where a human collided with the robot.*. The *openEASE framework* (Beetz et al. 2015) is equipped with the therefore required reasoning mechanisms, the visualization techniques, and the tool-chain to query episodic memories of robotic manipulation accordingly.

OpenEASE is here used to close the cognitive-control loop for wiping actions. The robot telemetry data (i.e. motions, torques, and forces) is therefore augmented with the semantic information about desired and undesired contacts, effects to the particle model, and the resulting task performance. These *narrative-enabled episodic memories* are provided to the openEASE framework to make it accessible on a semantic level.

The semantic augmentation of logged data streams allows to query big data based on narratives that are grounded in an ontology. In the example at hand, researchers can query multiple episodic memories of compliant manipulation tasks at once, by formulating relatively simple queries based on symbols related to physical interaction, such as contact or collision. Ideally, this narratives are generated during runtime by the control program of the robot (Beetz et al. 2016). This is possible if all states and transitions are fully observable as it is common for state machines and symbolic planners. In the example at hand, this would mean that the procedure (e.g. a wiping action) logs its current state based on planned events, such as the expected contact with the table surface. However, as argued in the introduction, this is often not possible due to difficult lightning conditions, imprecise localization, or external disturbances that are unpredictable. As a result, the logged semantics do not correspond to the logged sensor stream. Moreover, many systems lack symbolic task information, such as teleoperated robots for space operations or minimal invasive surgery. To circumvent this issues, the sensor stream is annotated offline (Haidu and Beetz 2016), based on the reasoning mechanisms presented in Sect. 7.1. As a result, the openEASE system can be queried for all collision events that occurred during a trial as seen in Listing 7.1.

Listing 7.1 Prolog query to request and visualize all collision events from openEASE.

```
owl_individual_of(C, semco:'Collision'),
     occurs(C, [S, E]),
     M is 0.5 * (S + E),
     show(trajectory('sponge'),
     interval(S, E, dt(0.5))),
     show(justin:'justin_robot1', M).
```

The query language in openEASE is based on *Prolog* (Sterling and Shapiro 1994). In the example at hand, the episodic memory of a particular trial is queried for all collisions `C`. Given the start time `S` and the end time `E`, one can visualize the sponge trajectory at the entire interval of the collision. Additionally, one can calculate the intermediate time frame `M` to show the configuration of the robot in the middle of the event. The result is visualized in Fig. 7.18. This relatively short and simple query reveals that the robot was interrupted in the center of the chopping board, which leads to a significantly decreased performance. This information would otherwise be unavailable from feedforward data and hard to identify based on purely numeric telemetry.

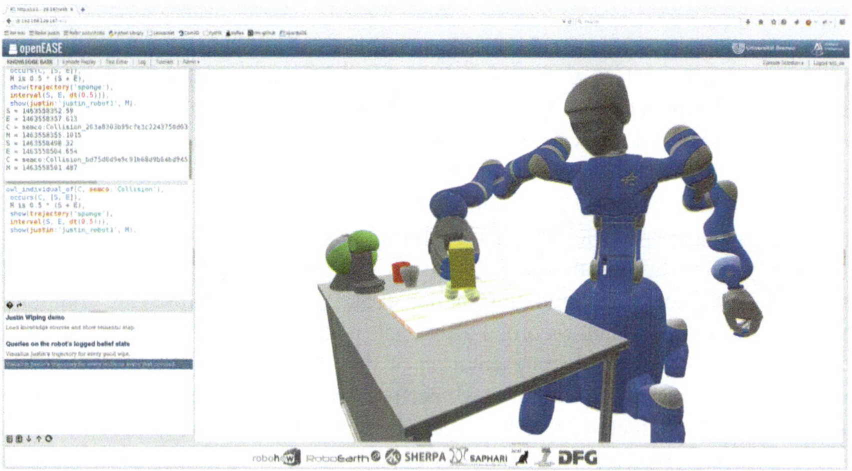

Fig. 7.18 The web interface of openEASE visualizing all trajectories of the sponge where a collision occurred. The yellow trajectory displays one of several collision events

The example at hand is just a first step toward more powerful queries. To give an illustration, a thoughtful post-treatment of raw sensor streams may hint toward design flaws of the motion planning algorithms or the applied control strategy. On the other hand, collecting data from multiple robots can be used to assess the performance of a single robot w. r. t. the data of an entire assembly line. Furthermore, one may be able to identify failure states before they appear by means of predictive computation. In conclusion, the possibility to query big data based on the abstract nature of a task is a powerful tool for robotics and AI researchers.

7.3 Related Work

Wiping motions are often considered as fundamental part of cleaning actions and have therefore been investigated in robotics research in the recent past to some extend. The most related articles are listed as follows. Gams et al. (2014) investigate cleaning motions from a learning point of view. They exploit the compliance of a LWR III robot to adapt to unknown surface geometries and modify the cleaning motion w. r. t. physical contact introduced by a human tutor. This way a human can directly modify the periodic cleaning motions. However, there is no particular goal specified as the wiping motions are considered as prototypical actions. Hess et al. conducted research on robotic cleaning in a series of papers (Hess et al. 2011, 2014). They investigate cleaning as a path coverage problem for robotic manipulators and vacuum robots. In Hess et al. (2011) they learn the effect of a vacuum cleaner moving along a planar surface by utilizing visual feedback based on color segmentation. The robot can enhance the task execution for future trials as it generates motions that cover only the dirty areas. A similar approach is used to plan optimal motions for a vacuuming robot (Hess et al. 2014). Based on a discretized dirt distribution grid model the robot plans the motions to clean the floor most efficiently. This model is learned by utilizing a dirt sensor that measures the impact of dirt particles. In both articles it is implicitly assumed that dirt is absorbed upon contact. Martínez et al. (2015) investigate planning for robotic cleaning by wiping with a sponge under the assumption that the particles are pushed upon contact. Additionally, the authors propose to represent dirt accumulations as ellipses to have them accessible as semantic predicates for automated planning (Ghallab et al. 2004). The ellipses are computed based on color image data recorded after each wiping motion. Similarly, Do et al. (2014) utilize a perceptual representation of dirt distributions on a target surface to derive a scalar value to rate the task performance w. r. t. different object properties and action parameters. The listed articles provide some strong arguments for visual perception which is a natural selection. However, even though these research groups make implicit assumptions of the wiping effect, there is no underlying model available that could be used to predict the actual task performance without visual validation as physical contact is not explicitly modeled. Accordingly, the robots are able to interpret visually perceivable pollutions, but they cannot make assumptions on the task performance from haptic feedback.

In contrast, there has been some work on contact effect modeling for wiping motions and effect estimation for robotic manipulation tasks in general. Kunze et al. (2011) utilized a simplified process model based on particle simulations to infer the effect of tools interacting with their environment. In particular, they simulate the effect of a sponge absorbing liquids in contact. This way, a qualitative effect inference can be conducted based on the absorbed and leftover water particles. Winkler and Beetz (2015) maintain expectation about the outcome of planned manipulations in pick-and-place scenarios. Based on observation of relevant task parameters (e. g. gripper forces during object transitions) a robot can learn when an action was successfully executed or failed. Pastor et al. (2011) propose to learn motor skills in form of Dynamic Movement Primitives. Additionally they predict the task outcome of the manipulation tasks based on statistical hypothesis testing w. r. t. low-level sensor streams which include force information recorded with fingertip pressure sensors. The proposed approach enables the robot to predict failure situations online. Obviously, contact force is also of interest for the contact-rich task of cleaning. However, there has not yet been any efforts to qualitatively model the effect of wiping motions on a high level of abstraction based on low-level control feedback.

7.4 Summary

The curiosity to explore, understand, and interpret the world is a remarkable intellectual virtue that makes humans incomparable to any other known species. Cognitive science can provide clues on special cases of human intelligence (Kawato 1999; Gibson 1962; Flanagan et al. 2006), yet, unfortunately it is unable to provide an answer to the big picture. Nonetheless, creating artificial cognitive capabilities that mirror human-like intelligence is the ultimate challenge for AI researchers. Prominent examples are the IBM Watson system (Ferrucci 2012), and AlphaGo by DeepMind (Silver et al. 2016). However, these systems solve purely computational problems based on big data. This issues cannot be compared to the challenges a robot faces in a real world. Especially, the interpretation of the effects resulting from compliant physical interaction possess a non-trivial difficulty. Similarly to the quest for an answer of cognitive science on human intelligence, it is impossible for this manuscript to provide a universal answer to the high-level interpretation of physical interaction tasks. However, this chapter proposed an approach to interpret wiping motions on a semantic level. In particular, the particle distribution model introduced in Chap. 4 was used to estimate the effect of robotic cleaning actions w. r. t. haptic feedback information. A log-likelihood based contact model was utilized to simulate different tool-medium-surface combinations. The proposed method is capable of detecting failure situations occurring from bad localization and human intervention. As a result, the robot was able to question its own actions and plan additional wiping motions to accomplish the goal. At last, the openEASE framework was used to query episodic memories based on the symbolic nature of the task. All things considered, the findings of this chapter provide the means for a compliant robot to understand its interaction with the physical world in the context of wiping tasks. The next chapter applies the developed reasoning methods in a relevant scenario, i. e. the robotic exploration of distant planets.

References

Albu-Schäffer, Alin, Sami Haddadin, Christian Ott, Andreas Stemmer, Thomas Wimböck, and Gerd Hirzinger. 2007a. The DLR Lightweight Robot: Design and Control Concepts for Robots in Human Environments. *Industrial Robot: An International Journal* 34 (5): 376–385.

Beetz, Michael, Moritz Tenorth, and Jan Winkler. 2015. Open-EASE-A Knowledge Processing Service for Robots and Robotics/AI Researchers. In *IEEE International Conference on Robotics and Automation (ICRA)*, 983–1990.

Beetz, Michael, Daniel Beßler, Jan Winkler, Jan-H. Worch, Ferenc Balint-Benczedi, Georg Bartels, Aude Billard, Asil K. Bozcuoglu, Zhou Fang, Nadia Figueroa, Andrei Haidu, Hagen Langer, Alexis Maldonado, Ana-Lucia Pais, Moritz Tenorth, and Thiemo Wiedemeyer. 2016. Open Robotics Research Using Web-based Knowledge Services. In *Proceedings of the International Conference on Robotics and Automation (ICRA)*, 5380–5387.

Denei, Simone, Fulvio Mastrogiovanni, and Giorgio Cannata. 2015. Towards the Creation of Tactile Maps for Robots and Their use in Robot Contact Motion Control. *Robotics and Autonomous Systems* 63: 293–308.

Do, Martin, Julian Schill, Johannes Ernesti, and Tamim Asfour. 2014. Learn to Wipe: A Case Study of Structural Bootstrapping from Sensorimotor Experience. In *Proceedings of the IEEE International Conference on Robotics and Automation (ICRA)*, 1858–1864.

Ferrucci, David A. 2012. Introduction to "This is Watson". *IBM Journal of Research and Development*, 56 (3.4): 1–1.

Fischler, Martin A, and Robert C Bolles. 1981. Random Sample Consensus: A Paradigm for Model Fitting with Applications to Image Analysis and Automated Cartography. *Communications of the ACM*, 24 (6): 381–395.

Flanagan, J Randall, Miles C Bowman, and Roland S Johansson. 2006. Control Strategies in Object Manipulation Tasks. *Current Opinion in Neurobiology*, 16 (6): 650–659.

Gams, Andrej, Tadej Petric, Bojan Nemec, and Ales Ude. 2014. Learning and Adaptation of Periodic Motion Primitives Based on Force Feedback and Human Coaching Interaction. In *IEEE-RAS International Conference on Humanoid Robots (ICHR)*, 166–171.

Ghallab, Malik, Dana Nau, and Paolo Traverso. 2004. *Automated Planning: Theory and Practice*. Morgan Kaufmann.

Gibson, James J. 1962. Observations on Active Touch. *Psychological Review*, 69 (6): 477.

Gibson, James J. 1966. The Senses Considered as Perceptual Systems.

Haidu, Andrei, and Michael Beetz. 2016. Action Recognition and Interpretation from Virtual Demonstrations. In *Proceedings of the IEEE/RSJ International Conference on Intelligent Robots and Systems (IROS)*, 2833–2838. IEEE.

Hazara, Murtaza, and Ville Kyrki. 2016. Reinforcement Learning for Improving Imitated In-Contact Skills. In *Proceedings of the IEEE-RAS 16th International Conference on Humanoid Robots (ICHR)*, 194–201. IEEE.

Hess, Jürgen, Jürgen Sturm, and Wolfram Burgard. 2011. Learning the State Transition Model to Efficiently Clean Surfaces with Mobile Manipulation Robots. In *Proceedings of the Workshop on Manipulation under Uncertainty at the IEEE International Conference on Robotics and Automation (ICRA)*.

Hess, Jürgen, Maximilian Beinhofer, and Wolfram Burgard. 2014. A Probabilistic Approach to High-Confidence Cleaning Guarantees for Low-Cost Cleaning Robots. In *Proceedings of the IEEE International Conference on Robotics and Automation (ICRA)*, 5600–5605.

Hirzinger, Gerd, Norbert Sporer, Alin Albu-Schäffer, M Hähnle, Rainer Krenn, A Pascucci, and Manfred Schedl. 2002. DLR's Torque-controlled Light Weight Robot III-Are We Reaching the Technological Limits Now? In *Proceedings of the IEEE International Conference on Robotics and Automation (ICRA)*, vol. 2, pp. 1710–1716. IEEE.

Hogan, Neville. 1987. Stable Execution of Contact Tasks using Impedance Control. In *Proceedings of the IEEE International Conference on Robotics and Automation*, vol. 4, pp. 1047–1054.

Kawato, Mitsuo. 1999. Internal Models for Motor Control and Trajectory Planning. *Current Opinion in Neurobiology* 9 (6): 718–727.

Kunze, Lars, Mihai Emanuel Dolha, Emitza Guzman, and Michael Beetz. Simulation-based Temporal Projection of Everyday Robot Object Manipulation. In *Proceedings of the International Conference on Autonomous Agents and Multiagent Systems (AAMAS)*, 107–114.

Leidner, Daniel, and Michael Beetz. 2016. Inferring the Effects of Wiping Motions Based on Haptic Perception. In *Proceedings of the IEEE/RAS International Conference on Humanoid Robots (ICHR)*, 461–468.

Martínez, David, Guillem Alenya, and Carme Torras. 2015. Planning Robot Manipulation to Clean Planar Surfaces. *Engineering Applications of Artificial Intelligence* 39: 23–32.

Norman, Donald A, and Tim Shallice. 1980. Attention to Action: Willed and Automatic Control of Behavior. Technical report, DTIC Document.

Pastor, Peter, Mrinal Kalakrishnan, Sachin Chitta, Evangelos Theodorou, and Stefan Schaal. 2011. Skill Learning and Task Outcome Prediction for Mmanipulation. In *Proceedings of the IEEE International Conference on Robotics and Automation (ICRA)*, 3828–3834. IEEE.

Silver, David, Aja Huang, Chris J Maddison, Arthur Guez, Laurent Sifre, George Van Den Driessche, Julian Schrittwieser, Ioannis Antonoglou, Veda Panneershelvam, Marc Lanctot, et al. 2016. Mastering the Game of Go with Deep Neural Networks and Tree Search. *Nature*, 529 (7587): 484–489.

Šmilauer, Václav, Emanuele Catalano, Bruno Chareyre, Sergei Dorofeenko, Jerome Duriez, Anton Gladky, Janek Kozicki, Chiara Modenese, Luc Scholtès, Luc Sibille, et al. 2010. Yade Reference Documentation. *Yade Documentation*, 474.

Sterling, Leon, and Ehud Y Shapiro. 1994. *The Art of Prolog: Advanced Programming Techniques*. MIT Press.

Winkler, Jan, and Michael Beetz. 2015. Robot Action Plans that Form and Maintain Expectations. In *Proceedings of the IEEE/RSJ International Conference on Intelligent Robots and Systems (IROS)*, 5174–5180.

Chapter 8
Applied Intelligent Physical Compliance

This chapter shall extend the horizon of this manuscript beyond daily household chores and gives an outlook to the possibilities that emerge with the availability of cognitive robots with human-like manipulation capabilities. In particular, this chapter applies the developed reasoning methods to an actual space robotics mission. Section 8.1 shows how the concepts contribute to the METERON SUPVIS Justin experiment. The Action Template approach is exploited to develop a supervised autonomy user interface (UI) to command the robot as described in Sect. 8.2. Eventually, it is shown how the concept of Intelligent Physical Compliance is used to solve the mission objectives in Sect. 8.3.

The findings of this chapter are based on the scientific output of the METERON SUPVIS Justin experiment. The milestones of the experiment are documented in (Lii et al. 2015a, b, 2017). The interface is based on the supervised autonomy concept initially developed in Birkenkampf (2013), Birkenkampf et al. (2014) and specialized to the needs of planetary exploration in Leidner et al. (2014a), Birkenkampf et al. (2017), Leidner et al. (2017).

8.1 The METERON SUPVIS Justin Experiment

The general deployment of cognition-enabled service robots remains a challenging task for most everyday domains. While compliant industrial robots are already utilized to execute delicate, recurring manufacturing tasks (e.g. screwing and bolting), service robots are not yet immediately able to solve everyday manipulation task beyond vacuuming the floor. However, there are several application fields where robots with advanced cognitive capabilities are under consideration for use as productive tools, such as disaster response, handling of hazardous material, and space

D. S. Leidner, *Cognitive Reasoning for Compliant Robot Manipulation*, Springer Tracts in Advanced Robotics 127,
https://doi.org/10.1007/978-3-030-04858-7_8

exploration. Especially the space exploration sector is eager to utilize robots for *On-Orbit Servicing (OOS)* missions (e.g. satellite maintenance and space debris removal Pelton 2015), assist astronauts in *Extra Vehicular Activities (EVA)* (Diftler et al. 2011), and the exploration of distant planets such as Mars (Blake et al. 2013).

The exploration of Mars has received tremendous attention after NASA presented their plans to send a human to Mars in the early 2030s. However, no human can survive on Mars without a habitat and the infrastructure to provide energy, breathable air, water, and food. Yet, habitats and infrastructure cannot be set up without being on-site. A possible solution to this issue is the deployment of robots to pave the ground for humankind. However, this scenario highlights the necessity to advance the autonomy of teleoperated rovers toward *intelligent space robot assistants*. Robots on distant planets can no longer be safely and efficiently teleoperated by means of direct control, since the distance to the target location is too far to command continuous workspace motions. To give an illustration, the distance from Earth to Mars varies from 0.372 AU to 2.683 AU.[1] This means that it takes up to 22.27 min at the speed of light c to transmit a radio signal from Earth to a robot on the Mars surface. In return, this doubles the time to receive an answer of the robot to about 45 min. Accordingly, autonomous operation is the key to effective robot operation on Mars. Suitable methods to instruct a robot semantically by means of goal-oriented high-level commands are unavoidable. In addition, it is mandatory for such a robot to exhibit compliant behavior for the sake of safety and w.r.t. the numerous tasks in unknown environments and the delicate equipment that requires careful handling.

Based on this demands, the *Multi-Purpose End-To-End Robotic Operation Network* Project, or METERON (Schiele 2011), aims to develop a network infrastructure to teleoperate robots in future space exploration missions. It is an expanding and evolving suite of experiments for technology demonstration that is initiated by the *European Space Agency (ESA)*. METERON is carried out in partnership with NASA, Roscosmos, and DLR. Led by DLR in partnership with ESA, SUPVIS Justin is one of the experiments in the METERON suite. To avoid the high latencies illustrated in the introduction of this section, the project envisions a permanently manned space station in the orbit of Mars to control robots with signal delays of 800 ms and above. The *International Space Station (ISS)* is used as research vessel to provide answers about the technology requirements for robotic space exploration. The main goal of the project is to validate autonomous and telerobotic operation from space to ground. In particular, several robots located on Earth are controlled from an astronaut inside the ISS. The control modalities vary based on the scope of the experiment. The *Haptics-1* experiment studies the effects of micro-gravity on psycho-motor performance w.r.t. haptic teleoperation using a force-feedback joystick (Schiele et al. 2016). *Haptics-2* builds on these findings and uses the force-feedback joystick on the station to operate a second joystick on ground (Krueger and Schiele 2015). The *Interact* experiment uses haptic teleoperation paired with discrete operational space commands and state-of-the-art software concepts to control the Interact Centaur robot to solve

[1]One *Astronomical Unit (AU)* refers to the mean distance from the center of the Sun to the center of the Earth, which is about 149.6 million km. It takes 8.3 min to travel 1 AU at the speed of light.

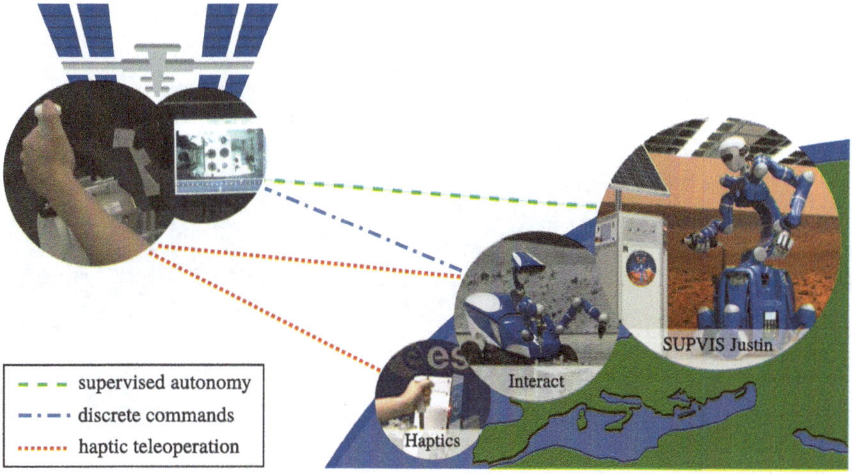

Fig. 8.1 Overview of related experiments in the METERON project. The Haptics experiments validate bilateral haptic teleoperation. Interact uses a mixture of telepresence and basic high-level commands to command a mobile robot. In contrast, the SUPVIS Justin experiment uses a full set of advanced high-level commands to supervise the actions of a robot

a peg-in-hole task (Schiele 2015). The *SUPVIS Justin* experiment takes advantage of the AI-reasoning methods developed in this work, to command Rollin' Justin via supervised autonomy commands on a high level of abstraction. The robot uses its local intelligence to solve the tasks autonomously based on the current mission context (Lii et al. 2015a, b). An overview of the project is given with Fig. 8.1. It shows the main input devices, i.e. the force-feedback joystick and a tablet computer, as well as three of the main target platforms for the individual experiments.

Depending on the input method and the targeted signal delay, different requirements to the communication channel are defined. In particular, haptic teleoperation requires low latency to guarantee stability of the system, since haptic feedback becomes unstable with increasing latency. Direct control requires sufficiently high bandwidth such that a human can operate the robot in real-time. Additionally, a timely video stream is needed to follow the procedure visually. On the other hand, supervised autonomy can be applied in case of high latency and low bandwidth. This is due to the fact that an operator has to command high-level goals only. Once a task is commanded, the robot schedules a sequence of actions and executes them autonomously. In the mean time, there is no communication with the operator required. The last scenario is subject of the SUPVIS Justin experiment. It represents the main focus of this chapter.

The SUPVIS Justin experiment is spearheaded by the Institute of Robotics and Mechatronics at DLR. The mission scenario is that of a Mars colony. Rollin' Justin is stationed in a simulated extraterrestrial solar power plant on Mars, the *SOLar Farm EXperimental Environment, or SOLEX* facility (located on Earth). This facility consists of several *Solar Panel Units (SPU)* that have to be surveyed and maintained

Fig. 8.2 The image shows the astronauts Paolo Nespoli and Jack Fisher during the first experiment session from ISS on August 25, 2017. Paolo Nespoli (in red on the monitor) trained Jack Fisher (in blue on the monitor) and Randy Bresnik (not shown) spontaneously in flight. The intuitive interface concept allowed the additional astronauts to control Rollin' Justin without prior experience

by the aid of the robot. The SPUs consist of a modular base and a solar panel mounted on a passive pan-tilt unit. The base provides the control panel in form of a switch board. A data port allows the robot to physically connect to the SPUs by means of a *Data Interface Probe (DIP)*. All of these components have to be handled carefully, such that compliant contact behavior becomes unavoidable. Action Templates have been developed for each action following the concept of Intelligent Physical Compliance to enable the robot to interact with the SPUs.

Three experiment session were conducted on August 25, 2017, March 02, 2018, and August 17, 2018. The first crew session is shown in Fig. 8.2. ESA astronaut Paolo Nespoli, NASA astronaut Jack Fisher, and NASA astronaut Randy Bresnik performed four trials in total. The astronauts on board the ISS control Rollin' Justin via a tablet computer application. They are provided with a video stream of the head mounted cameras and allowed to freely change the joint angles of the robot neck to change the viewpoint. Furthermore, the astronauts are able to navigate the robot toward a specific goal on a predefined map. Eventually, the astronaut can select high-level goals to command the robot to interact with its environment. The supervised control concept is detailed in the following section.

8.2 A Supervised Autonomy UI Based on Action Templates

Operating robots from a distance is always difficult as it is hard for an operator to focus on the task to be solved, rather then on the robot that is controlled. It is particularly hard to interpret the world based on the capabilities of the commanded robot. That is, an operator may tend to overestimate the workspace of the robot, it may be unclear which actions it can perform, and it is difficult to interpret the effects of the executed action. Moreover, even though astronauts are exceptionally skilled and trained for their mission, the cognitive capabilities are negatively impaired during human space flight in zero gravity (Manzey and Lorenz 1998). For this reason, it is often desired to control robots based on a direct basis, i.e. by means of force-feedback devices such as joysticks or exoskeletons. However, as previously stated, this requires low-latency communication, which is not available within the SUPVIS Justin experiment. Accordingly, this section proposes to abstract away from the direct control of robotic components (e. g. arms for manipulation or the base for navigation) toward a task-centric UI. As already stated in the earlier chapters, it is natural for humans to connect action possibilities with the objects that are perceived (see the concept on affordances in Sect. 4.1). With this in mind, an intuitive UI to teleoperate robots by means of supervised autonomy is developed. The application is designed for the use on a tablet computer. This design choice is based on the research conducted by (Rouanet et al. 2013) on the remote operation of robotic systems, where it is emphasized that touch screens provide intuitive input possibilities and visual feedback likewise. The SUPVIS Justin experiments are conducted on a Dell Latitude 10 tablet PC, which has been upmassed to the ISS with the *Automated Transfer Vehicle ATV-5* in July 2014 for the Haptics-1 experiment (Schiele et al. 2016).

Using the SUPVIS Justin UI, an astronaut can initially get an overview of the surrounding of the robot by means of an interactive map generated based on the simulated world state of the robot. This is especially useful due to the fact that the *Mars Reconnaissance Orbiter (MRO)*, which may be able to provide images of the real area of operation, travels in a polar orbit to observe the entire Mars surface, rather than a geostationary orbit to cover a distinct area. A particular map segment is visualized in the screen shot shown in Fig. 8.3. This semantic map provides the astronaut not only the information where the robot is located w.r.t. the SPUs, but also the possibility to navigate directly to a certain object by selecting it on the screen. As described in Algorithm 5.3 the robot would autonomously travel to the target. This way the astronaut can abstract away from directly controlling the base of the robot and focus on other tasks in the meantime.

The main view of the UI is shown in Fig. 8.4. The snapshot of the UI is taken shortly after the still frame shown in Fig. 8.2. In this case, the main canvas of the UI presents a live video feed of the high-resolution camera mounted on the head of the robot. The camera view can be changed by means of the pan-tilt sliders on the bottom and the left, respectively. Furthermore, it is also possible to click on the camera view to move the head directly to a desired goal position, e. g. to closer inspect a salient area in the video. Yet, the most compelling feature of the UI is the possibility to interact with

Fig. 8.3 Navigation view of the supervised autonomy UI. The semantic map is based on the internal world state of the robot. It is possible to move toward a freely selectable goal position (orange marker) or toward a particular object

the environment. In particular, the objects that are known to the robot are augmented to the video stream as blue overlay. Once an object is selected (orange overlay), the UI lists all context specific interaction possibilities (right) to the astronaut. Each of these actions is implemented as a specific Action Template. Only the actions that are currently appropriate are authorized for execution, where symbolic and geometric filter criteria apply. Based on these actions, the astronaut has to achieve the mission objectives listed on the left. The pseudo code in Algorithm 8.1 explains the filter procedure.

The algorithm reviews all Action Templates α provided by the class of the selected object $\mathcal{O}$ and its parent classes. To cover all action possibilities, the world state is recursively parsed for all available parameter combinations by the *RecursivelyResolveParameters* function. It substitutes all action parameters with matching objects currently in the world state. The resulting list of available actions α_r is validated symbolically and geometrically. Symbolic filters mainly consider the length of the anticipated action sequences denoted as γ. If the effects $\boldsymbol{e}$ for an action are already a subset of the predicates in the world state $\boldsymbol{\omega}$, the action is obsolete and therefore not presented to the astronaut ($\gamma \leftarrow 0$). If the preconditions $\boldsymbol{p}$ are immediately reachable as subset in the world state $\boldsymbol{\omega}$, it requires only one action to solve the problem ($\gamma \leftarrow 1$). Otherwise, the symbolic planner has to calculate the shortest available sequence of actions. Depending on the objectives, mission control may refuse to authorize action sequences of a certain length, since failure likelihood increases as the traveled joint path accumulates. Geometric filters may refuse actions that are too far away which forces the astronaut to explore the map using the navigation view.

Fig. 8.4 The central area of the manipulation view shows the augmented video stream of the robot. After an object is selected, a list of possible actions is shown

---

Algorithm 8.1: Resolving action parameters and apply filters.

Input: The object of interest $\mathcal{O}$.
Output: The list of authorized actions $\mathcal{A}$.

$\mathcal{A} \leftarrow$ List()
foreach $\alpha \in$ *GetAllActionTemplates($\mathcal{O}$)* **do**
  foreach $\alpha_r \in$ *RecursivelyResolveParameters(α)* **do**
    $\boldsymbol{p} \leftarrow$ GetPreconditions(α_r)
    $\boldsymbol{e} \leftarrow$ GetEffects(α_r)
    if $\boldsymbol{e} \subset \omega$ **then**
      $\gamma \leftarrow 0$
    else if $\boldsymbol{p} \subset \omega$ **then**
      $\gamma \leftarrow 1$
    else
      $\mathcal{T}_{sym} \leftarrow$ SymbolicPlan($\boldsymbol{e}$)
      $\gamma \leftarrow$ Length($\mathcal{T}_{sym}$)
    if *ApplySymbolicFilters(α_r, γ)* **then**
      continue
    if *ApplyGeometricFilters(α_r)* **then**
      continue
    $\mathcal{A}^\frown \langle \alpha_r, \gamma \rangle$
return $\mathcal{A}$

8.3 Mission Objectives and Scientific Goals

The mission in the SUPVIS Justin experiment simulates a future space exploration mission, in which a crew of astronauts orbits Mars on board a space craft (simulated by the ISS orbiting Earth). The aim is to setup habitats and infrastructure for manned Mars exploration, by means of remote controlled intelligent service robots. Accordingly, the communication time to the robot varies from 800 ms up to a few minutes. However, communication to Earth takes up to 45 min such that the operator has to make its own decisions without the aid of a robotics specialist at mission control. When the astronaut starts the experiment, he/she is confronted with a notification about an incident in the SOLEX facility. The astronaut receives the mission objective to isolate the malfunctioned SPU, inspect it, and restore it to working order.

The scientific goals of the SUPVIS Justin experiment are defined in the experiment protocols. The first protocol, studies the ability to survey and inspect a remote location by means of a robot avatar. This protocol introduces the astronaut to the touchscreen application and the UI. It can be considered a tutorial that has to be accomplished to qualify for the main experiment. Utilizing the head mounted camera of the robot, the astronaut can survey the environment and record suspicious objects or events. After the broken SPU is identified, the robot is commanded to navigate toward it.

The second protocol investigates the ability of an operator in zero gravity to focus on a complex manipulation task executed by a remote controlled service robot. It is thereby of particular interest to understand and quantify the boundaries of complexity at which the cognitive load becomes too high for an astronaut to effectively carry out the task. This is especially crucial as the chosen interface does not provide a full immersiveness and the potential end-use application aims to control multiple robots at different locations at once. For this reason, the particular scenario in the second protocol only specifies that the astronaut has to recover the failure state of the SPU, but does not provide concrete information on the cause of the problem. The aim is to provide a more open-ended problem, as astronauts would encounter in a real repair mission. Accordingly, the astronaut has to explore the cause of failure on its own based on the actions provided by the available objects. For this, the robot would be commanded to download the status information of the SPU by connecting the DIP. Possible failure situations may include simple software errors, a misaligned solar panel, or low voltage of the system due to contaminated solar arrays. All problems require compliant behavior to interact with the reset switch, rotate the panel mounting, or clean the solar panel. To underline the scope of this work it is hereinafter discussed how the methods developed in this manuscript contribute to solve the last-mentioned task, i.e. skimming Martian dust and regolith off the solar panel.

As already mentioned, all actions provided to the astronaut are developed by means of Action Templates. The skimming action to clean the solar panel is no exception. It is the same as the action introduced in Sect. 5.3.1. In the example at hand, the action is parameterized based on the properties of the solar panel, a rubber wiper, and the dirt. The output of the reasoning procedure to plan the action is visualized as blue path (cf. Chap. 5). The haptic interpretation of the wiping motion is

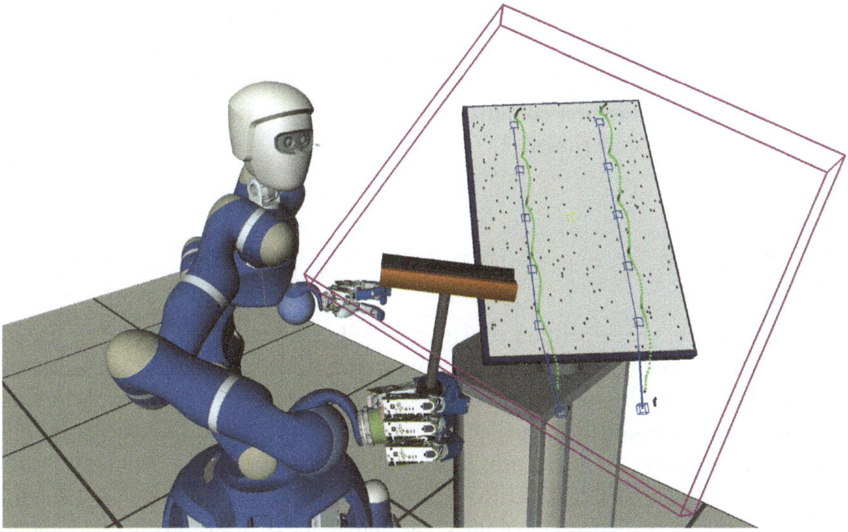

Fig. 8.5 Applied Intelligent Physical Compliance in the SUPVIS Justin experiment

Fig. 8.6 Rollin Justin cleaning the SPU as response to the identified failure status

visualized, where green path segments indicate the contact motion and a purple box highlights the estimated target surface, as seen in Fig. 8.5 (cf. Chap. 7). Eventually, the real execution of the task is shown in Fig. 8.6.

This final experiment proves the relevance of the research conducted in this work. As part of the technology demonstration planned for the METERON project, the concept of Intelligent Physical Compliance serves as important key technology to realize the SUPVIS Justin experiment. Moreover, the possibility to exploit Action Templates as part of the supervised autonomy UI shows the far-reaching consequences of this particular method to represent robot actions. In fact, this concept makes Rollin' Justin the first cognitive robot to co-operate with an astronaut in an actual space robotics experiment.

8.4 Summary

The exploration of space is often considered the human race's final frontier. However, it will probably be the first frontier for cognitive service robots. These robots are envisaged as semi-autonomous co-workers deployed to lighten the workload of astronauts in cumbersome, expensive, and dangerous situations. In view of this development, this chapter issued a prospect on the technology requirements for future service robots. As part of the METERON project, it was shown how compliant manipulation and intelligent decision making contribute to the SUPVIS Justin experiment. The representations and reasoning mechanisms presented in the earlier chapter were applied to a realistic scenario within the context of teleoperated robotic space exploration. In particular, it was described how Action Templates can be utilized to share task information with an astronaut and how Intelligent Physical Compliance provides an promising concept to realize tasks in extraterrestrial environments at the example of a solar panel cleaning task. In conclusion, this chapter extends the boundaries of this work toward the real world application of the developed methods.

References

Birkenkampf, Peter. 2013. Entwicklung Eines Human-Robot Interfaces Zur Visualisierung Interner Weltzustände und Steuerung potentieller Manipulationsfähigkeiteneines semiautonomen Roboters. Master's thesis, Technische Universität München.

Birkenkampf, Peter, Daniel Leidner, and Christoph Borst. 2014. A Knowledge-driven Shared Autonomy Human-robot Interface for Tablet Computers. In *Proceedings of the IEEE/RAS International Conference on Humanoid Robots (ICHR)*, pp. 152–159.

Blake, David F., Richard V. Morris, G. Kocurek, S.M. Morrison, Robert T. Downs, D. Bish, D.W. Ming, K.S. Edgett, D. Rubin, W. Goetz, et al. 2013. Curiosity at Gale Crater, Mars: Characterization and Analysis of the Rocknest Sand Shadow. *Science* 341 (6153): 1239505.

Birkenkampf, Peter, Daniel Leidner, and Neal Y. Lii. 2017. Ubiquitous User Interface Design for Space Robotic Operation. In *Proceedings of the 14th Symposium on Advanced Space Technologies for Robotics and Automation (ASTRA)*.

Diftler, Myron A, J.S. Mehling, Muhammad E Abdallah, Nicolaus A Radford, Lyndon B Bridgwater, Adam M Sanders, Roger Scott Askew, D Marty Linn, John D Yamokoski, FA Permenter, et al. 2011. Robonaut 2-the First Humanoid Robot in Space. In *Proceedings of the IEEE International Conference on Robotics and Automation (ICRA)*, pp. 2178–2183. IEEE.

Krueger, Thomas, and André Schiele. 2015. Preparations for the Haptics-2 Space Experiment onboard the International Space Station. In *Proceedings of the 13th Symposium on Advanced Space Technologies for Robotics and Automation (ASTRA)*.

Lii, Neal Y , Daniel Leidner, André Schiele, Peter Birkenkampf, Ralph Bayer, Benedikt Pleintinger, Andreas Meissner, and Andreas Balzer. 2015a. Imulating an Extraterrestrial Environment for Robotic Space Exploration: The METERON SUPVIS Justin Telerobotic Experiment and the SOLEX Proving Ground. In *Proceedings of the 13th Symposium on Advanced Space Technologies in Robotics and Automation (ASTRA)*.

Lii, Neal Y, Daniel Leidner, André Schiele, Peter Birkenkampf, Benedikt Pleintinger, and Ralph Bayer. 2015b. Command Robots from Orbit with Supervised Autonomy: An Introduction to the METERON SUPVIS Justin Telerobotic Experiment. In *Proceedings of the ACM/IEEE International Conference on Human-robots Interaction (HRI)*, pp. 53–54.

Lii, Neal Y., Daniel Leidner, Peter Birkenkampf, Benedikt Pleintinger, Ralph Bayer, and Thomas Krueger. 2017. Toward Scalable Intuitive Teleoperation of Robots for Space Deployment with the METERON SUPVIS Justin Experiment. In *Proceedings of the 14th Symposium on Advanced Space Technologies for Robotics and Automation (ASTRA)*.

Manzey, Dietrich, and Bernd Lorenz. 1998. Mental Performance During Short-term and Long-term Spaceflight. *Brain research reviews* 28 (1): 215–221.

Leidner, Daniel, Peter Birkenkampf, Neal Y. Lii, and Christoph Borst. 2014a. Enhancing Supervised Autonomy for Extraterrestrial Applications by Sharing Knowledge Between Humans and Robots. In *Proceedings of the Workshop on how to make best use of a Human Supervisor for Semi-autonomous Humanoid Operation at IEEE-RAS International Conference on Humanoid Robots (ICHR)*.

Leidner, Daniel, Neal Y. Lii, and Peter Birkenkampf. 2017. Context-Aware Mission Control for Astronaut-robot Collaboration. In *Proceedings of the 14th Symposium on Advanced Space Technologies for Robotics and Automation (ASTRA)*.

Pelton, Joseph N. 2015. Current Space Debris Remediation and On-orbit Servicing Initiatives. In *New Solutions for the Space Debris Problem*, pp. 11–29. Springer.

Rouanet, Pierre, Pierre-Yves Oudeyer, Fabien Danieau, and David Filliat. 2013. The Impact of Human-robot Interfaces on The Learning Of Visual Objects. *IEEE Transactions on Robotics* 29 (2): 525–541.

Schiele, André. 2011. METERON—Validating Orbit-to-ground Telerobotics Operations Technologies. In *Proceedings of the 11th Symposium on Advanced Space Technologies for Robotics and Automation (ASTRA)*.

Schiele, André, Manuel Aiple, Thomas Krueger, Frank van der Hulst, Stefan Kimmer, Jan Smisek, and Emiel den Exter. 2016. Haptics-1: Preliminary Results From the First Stiffness JND Identification Experiment in Space. In *Proceedings of the International Conference on Human Haptic Sensing and Touch Enabled Computer Applications*, pp. 13–22. Springer.

Schiele, André. 2015. Towards the Interact Space Experiment: Controlling an Outdoor Robot on Earth's Surface from Space. In *Proceedings of the 13th Symposium On Advanced Space Technologies for Robotics And Automation (ASTRA)*.

Chapter 9
Conclusions

The fundamental hypothesis that "*there is a need for intelligent physical compliance* was formulated in the introduction of this work, in order to account for the shortcomings of today's service robots. The hypothesis builds on the notion that particularly non-traditional robot tasks outside factory buildings demand compliant interaction capabilities in combination with advanced cognitive reasoning skills. However, this constitutes a need which is not even covered for application in laboratory environments. To overcome this issue, intelligent decision making strategies for compliant manipulation tasks have to be developed. That is, a robot has to process diverse task knowledge to *plan* effect-oriented actions, parameterize and *execute* these actions accordingly, and eventually *interpret* the outcome of the actions on a high-level of abstraction to estimate the task performance qualitatively. In order to facilitate this reasoning, suitable *representations* that incorporate the properties of physical compliant interaction have to be provided.

These aspects substantiate the hypothesis of a need for Intelligent Physical Compliance, which emerges from the hypotheses that "*there is a need for artificial intelligence*" and "*there is a need for physical compliance*" in order to fulfill society's "*need for service robots*". This motivation is well captured in the example of collecting leaves (cf. Fig. 1.1). The artist's conception illustrates Rollin' Justin deployed as a public service robot. It executes the recurring task of collecting leaves in a park area by the use of a rake. Among others, raking lines up with several tasks that are typically considered laborious and futile. Accordingly, the workload of many people could be relieved by delegating this task to a commercial service robot. However,

D. S. Leidner, *Cognitive Reasoning for Compliant Robot Manipulation*, Springer Tracts in Advanced Robotics 127,
https://doi.org/10.1007/978-3-030-04858-7_9

unlike most commercially available robots, Rollin' Justin possesses joint-torques sensors that allow for soft physical interaction with the environment. This allows it to guide the rake compliantly over the rugged cobblestone without loosing contact or damaging itself. Moreover, feedback-driven control strategies allow Rollin' Justin to constantly monitor the applied contact force and adapt to uncertainties. Nonetheless, compliant robots are not yet deployable as universal service robots. This is mainly due to the fact that the complexity of the problem grows with the capabilities of the robot. In this regard, stiff industrial robots are mainly utilized to execute rather simple actions as they are observed in pick-and-place tasks. These tasks can be represented by means of discrete changes to the world, i. e. the relocation of objects can be represented by homogeneous transformations. In contrast, the task of collecting leaves in a park requires a much richer process model. A robot must reason about the interaction of the rake and the leaves in an effect-oriented manner. It has to predict the effects in advance and estimate the real task outcome after execution. To achieve this, the robot requires qualitative representations of the effects, the actions, and the objects participating in the task execution. That is to say, the cognitive load for compliant manipulation tasks is much higher than it is for traditional robot operations, which explains the need for AI.

This final chapter summarizes the investigations conducted in this work in Sect. 9.1. General conclusions are drawn w. r. t. the research fields of robotics and AI in Sect. 9.2. Afterwards, the disclosed research questions are highlighted in Sect. 9.3. Finally, the hypothesis about a need for Intelligent Physical Compliance is discussed in the outlook given in Sect. 9.4.

9.1 Manuscript Summary

The combination of AI-reasoning methods and compliant robot control constitutes one key point for the developments conducted. In the introductory chapter, two hypotheses have been formulated to introduce the divergent viewpoints of the two research domains. The statements were intentionally formulated to be only partial correct, in order to illustrate the limited perspectives on the research question. However, the findings presented in this work confirm that the views on the problem are *not contradictory, but complementary* to each other such that they describe the whole problem in a deductive manner. Both statements are repeated and listed for direct comparison. This section summarizes the manuscript with reference to these statements and highlights the immediate contributions.

AI Research	Robotics Research
"First of all, the domain of the problem needs to be defined by means of an ontology. For example, one has to specify that a rake is a tool to collect leaves and it has to be carried by the robot to do so. The effect would be that the leaves are accumulated afterwards. Knowledge-based reasoning strategies have to be applied to infer that a rake has a certain probability to be picked up from a garden shed. A logic planner can schedule the actions accordingly."	*"Sweeping motions can be realized by means of whole-body impedance control. The tool motion would thereby be described by a moving virtual equilibrium point for the tool center of the rake. A virtual potential force aligns the tool w. r. t. the curvature of the target surface. Defining an appropriate Cartesian stiffness at the end-effector results in compliant contact behavior that allows to cope with environmental uncertainties and external disturbances."*

Starting from the high-level AI perspective, it was first stated that an ontology has to be defined in order to describe the domain of the problem. This has been done in form of the classifications conducted in Chap. 3. In order to generate a comprehensive classification of compliant manipulation tasks, the taxonomy was developed based on the nature of contact that can be observed during physical interaction between a hand/object system and the environment/target system. The classification terms describe the contact behavior during interaction, i. e. contact or no contact with the environment, the relevance of friction, the presence or absence of deformations, and the occurrence of penetrations. By further deepening the investigations, it appeared that especially wiping tasks are of great importance for service tasks in domestic and industrial settings. To that end, a detailed classification of wiping tasks was conducted based on the definition of tool-medium-surface tuples. This classification serves as the foundation of the reasoning methods developed. It included the prototypical tasks of absorbing, skimming, collecting, emitting, distributing, processing, scrubbing, grinding, and decomposing. Altogether, the taxonomy would provide a guideline for future researchers, on how to analyze compliant manipulation tasks and how to derive generic process models from this analysis.

With regard to the necessity of an ontology, it was further argued that a robot would require representations to describe objects, actions, motions, and effects of manipulation tasks. The manuscript covers this request in Chap. 4. Based the insights of cognitive science, it was derived that an object-centric representation is a natural approach to imitate human-like cognitive reasoning capabilities. As a consequence, all knowledge required to solve robotic manipulation tasks was arranged w. r. t. the hierarchical representation of functional object classes. Among others, this would involve geometric information such as CAD data, mass information, and visual features, but also symbolic knowledge including logic predicates and even handling instructions for robots in form of Action Templates developed in this work.

Action Templates represent a fundamental concept to describe the symbolic meaning of an action in combination with the geometric nature of the process model. They enable a robot to ground symbolic plans into executable robot commands by means of robot independent operations. In order to generate wiping motions based on generic process models in form of Action Templates, the so-called Semantic Directed Graphs (SDG) have been developed which represent Cartesian wiping motions on two-dimensional planes. In addition, a particle distribution model was developed to represent the effects of wiping actions. Altogether, the developed representations allow a robot to envision a problem mentally which is one of the most important skills for human-like problem solving.

The proposed representations provide the means to develop planning methods for everyday manipulation tasks. This complies with the final sentence of the AI statement, stating the need of knowledge-based reasoning strategies and logic planning to solve abstractly defined manipulation tasks. Chapter 5 is concerned with this issue and answered the elementary question on how high-level action definitions can be mapped on low-level robot manipulation skills, to eventually bridge the gap from AI research to robotics research. In particular, an object-centered hybrid reasoning framework was developed that utilizes Action Templates as fundamental building block for automated robotic manipulation. The framework is based on hybrid reasoning, i. e. the combination of symbolic planning and geometric planning. As such, the approach facilitates symbolic and geometric backtracking with integrated whole-body mobile manipulation planning capabilities based on reachability analysis. The framework has been extended toward effect-space planning methods for autonomous agents. That is, SDGs were combined with the particle distribution model to plan goal-oriented wiping motions based on abstract goal definitions for a certain medium. It was shown how the effect of these actions can be predicted qualitatively, in order to select the best available strategy in advance for a given scenario. The approach has been successfully applied to the tasks of absorbing, skimming, and collecting. The developed reasoning methods are based on the generic concept of effect-space planning, which is well suited to investigate automated manipulation in general.

The dependency and relationship between the two research domains became immediately evident with the second introductory statement, that is anticipated through system identification and problem analysis from the control engineering perspective. While it has been stated how sweeping motions can be realized, it has not been mentioned how these motions could be generated. In fact, it is often implicitly assumed that they are given to the closed-loop system based on the outcome of high-level planning methods. Nonetheless, the controller has to be parameterized in order to execute the motions correctly. This procedure was captured in Chap. 6 along with the execution of the motions. In particular, it was argued that complaint contact behavior would be best realized by means of hierarchical whole-body impedance control. This approach would allow to resolve the highly redundant null space of humanoid robots while it is additionally possible to satisfy secondary control tasks such as self-collision avoidance, singularity avoidance, and gravity compensation. To this end, an approach to integrate the whole-body controller of Rollin' Justin into the developed high-level reasoning framework has been formulated. The approach was validated in three elaborate experiments, namely, scrubbing a mug with a sponge, skimming detergent off a window, and collecting shards of a broken mug with a broom. The experiments revealed that it is not always straightforward to design the desired in-contact motion for a given scenario. However, it was confirmed that it would be possible to exploit the compliant contact behavior of the robot. Notably, it was recommended to design the actual tool motion as a sequence of translations only and introduce rotations by means of low rotational stiffness settings to comply with the curvature of the target.

The roboticist's argumentation continues with the noteworthy fact that the compliant contact behavior realized with impedance control would allow to cope with environmental uncertainties and external disturbances. While this is most desirable for human-robot collaboration, it is especially this feature that makes impedance control nondeterministic w. r. t. the task performance. Moreover, closed-loop system are typically unaware of the purpose of the generated motions, neither are they aware of the resulting outcome. To overcome this issue, an effect inference approach based on haptic feedback has been developed in Chap. 7. The approach is based on the analysis of episodic memories recorded during prior task executions. In particular, the recorded end-effector positions are referenced to the recorded Cartesian controller force to interpret nominal contact conditions, contact loss, and external disturbances. A probabilistic model has been applied to assess the quality based on the previously introduced particle model. In case of unpredicted disturbances, the robot is able to apply the reasoning methods introduced in Chap. 5 to plan additional motions in order to improve the task performance. In addition to the qualitative investigations, the tasks have been analyzed semantically by means of the openEASE framework, which is able to relate narratives to the logged sensor stream. Based on this data, this manuscript offers future researchers the possibility to investigate the conducted experiments to foster the development of novel reasoning methods.

Finally, the developed methods have been applied to a relevant scenario, i. e. robotic planetary exploration. While this type of task might be beyond the scope for most humans, space exploration was one of the first areas to utilize unmanned space crafts and rovers for the gain of knowledge. For this reason, the humanoid robot Rollin' Justin was initially developed as space robot assistant. In Chap. 8 it was outlined how the robot was utilized within the METERON project with the aim of advancing the remote control concepts for future robotic space exploration missions. In particular, it was shown how the concept of Action Templates contributes to the SUPVIS Justin experiment, in which Rollin' Justin was controlled by means of high-level commands issued by a tablet computer interface aboard the ISS. With this, Rollin' Justin could be considered *the first cognitive robot to co-operate with an astronaut*. As the mission of the robot was to maintain a solar power plant, the task of cleaning a solar panel was conducted for evaluation. The purpose of the last chapter was to emphasize the relevance of the conducted research w. r. t. the needs of society, and thus conclude the argumentation of this work.

9.2 Discussion

The aim of this work is the generic implementation of human-like cognitive reasoning skills to enable robots to reason about compliant physical interaction. While the main evaluation toward this goal has been conducted w. r. t. particular problem of wiping tasks, conclusions of general relevance can be drawn based on the obtained insights.

One key concept of this manuscript is the application of an object-centric perspective for the development of reasoning methods. This design choice is based on the mindset of affordances (Gibson 1986), which formulates a cognitive science approach to explain the subconscious mental perception of manipulation possibilities with objects. With reference to this, it can be related that object-centric examination of manipulation tasks has an immediate influence on the development of process models in form of Action Templates. Instead of focusing on the robot, a programmer would transfer the problem onto higher levels of abstractions (i. e. the interaction of objects), which leads to more generic solutions. Moreover, the representation of actions and their effects can be separated from kinematic reachability constraints (cf. Sect. 4.4), the description of process models can be designed independently of robot-specific capabilities (cf. Sect. 5.1.1), the controller can be natively parameterized in the Cartesian object-space (cf. Sect. 6.1.2), and the performance of the task execution can be related to object-specific spatial coordinates (cf. Sect. 7.1.2). In this context, objects serve as a well suited common ground to merge high-level reasoning methods and low-level control strategies.

During the elaboration of this manuscript, it has been observed repeatedly that impedance control poses a suitable approach to perform compliant contact. Evidence can be found in the successful execution of several wiping tasks as part of the experiments conducted with the humanoid robot Rollin' Justin. Impedance controlled motions enable a robot to adapt locally w. r. t. the target surface and even comply with uncertainties. The global motion planning problem can this way be reduced to rather simple motions consisting mainly of translations in the object-space. A well parameterized Cartesian stiffness can eventually introduce the necessary rotations to align with the curvature of the target. In this sense, one should exploit the local intelligence of the control level whenever possible, in order to reduce the load for the high-level reasoning framework.

Another major insight is based on the role of effect-space planning, which describes an attempt to represent effects such that planning methods can be developed to explicitly manipulate this representation. In the presented work, this approach has been applied to the domain of wiping tasks. In particular, a particle distribution model has been developed to enable the generic generation of wiping motions w. r. t. abstract goal definitions for the particles. The resulting approach is primarily independent of kinematic reasoning and thus applicable to any robot that is equipped with a manipulator. The proposed planning methods take the actual distribution of the particles into account in order to compute the most efficient tool motions. The quality of the planned motions can therefore be assessed for each step of a plan. This becomes especially valuable if the information is accessible via narrative-enabled episodic memories as it is possible with the openEASE framework. As a result it can be inferred that effect-space planning is of utmost importance for the generic development of robot manipulation skills.

Moreover, effect-space planning constitutes a necessary skill to master everyday manipulation tasks. This became evident ever since it turned out that the particle model is not only able to predict effects, but also to infer the effects of real world wiping motions. This feature would become one of the most compelling characteristics of the newly developed representation. In fact, the combination of planning methods and evaluation methods based on a shared effect representation is of great potential. In the particular case discussed in this work, it allows to assess the efficiency of wiping actions executed by the robot and improve them directly in the effect-space. This feature leads to a significant quality boost as the robot is made aware of its own performance. As a consequence, the robot may put the quality of its own motions into question, especially if it identifies a failure situation, such as the collision with a human. Generally speaking, it is recommended to develop effect representation in such a way that they incorporate both a forward model to predict effects and an inverse model to infer effects likewise.

Finally, it can be confirmed that haptic feedback is a valuable, yet underestimated source of information. It has been shown that a compliant robot can estimate the quality of wiping motions purely based on haptic perception, i.e. the combination of proprioception and contact force information (Gibson 1966). This is especially interesting for scenarios that lack sufficient visual perception due to bad lighting conditions or invisible materials such as liquids or small particles. In spite of the fact that vision is often considered to be the primary sense in robotics, it is highly recommended to fuse visual perception with haptic perception in order to get more reliable data. Accordingly, the application of haptic perception by means of torque sensors, force sensors, and tactile skin should be taken into account whenever possible.

9.3 Open Research Questions

The conducted research disclosed three major open challenges in the context of Intelligent Physical Compliance and robotic manipulation in general. At first, the proposed approach covers the symbol grounding problem for actions in a generalized form. However, a general solution to the symbol grounding problem for effects remains an open problem. This imbalance originates from the fact that actions can be described as a sequence of atomic operations, while it is hardly possible to describe an effect by a set of atomic effects. Moreover, purely symbolic effect descriptions cannot be evaluated in a generic fashion. As a result, one may have to examine each task family individually in order to derive locally generic effect representations, as it was done for wiping tasks in this work.

Second, it is difficult to model actions and effects in such a way that they match every possible instance of a problem. Accordingly, the model parameters for a particular class of tasks have to be fine-tuned for every single trial. However, this is not sufficient for the general deployment of future service robots. These robots will have to optimize model parameters online by means of machine learning algorithms in order to be truly advantageous. This is both desired for the forward model (i. e. the action model) as well as the inverse model (i. e. the effect model). This way, robots may be able to continually improve their effectiveness and eventually catch up with human skill level.

The third research question constitutes on of the most difficult open problems in robotic manipulation, i. e. the detection, isolation, and recovery of arbitrary failure states. This is particularly complicated as the possible failure states exceed the number of nominal states by far. It is unpredictable which kind of error may occur during a task execution. Moreover, there is no generic fallback strategy available to recover from arbitrary failure states. In the worst case a robot may have to replan from scratch or maybe even abort the task execution completely if the task becomes unfeasible. It remains an open question how the required perception and reasoning methods can be integrated into a generic robot control program.

9.4 Outlook

Considering the short-term goals, it is of great relevance to generalize the developed approaches for other manipulation tasks that require both physical compliance and intelligent behavior. One particular example is found in cutting vegetables and other food products with a knife as it was presented by Lenz et al. (2015). The contact force and the cutting motion show a relation that is comparable to that of wiping actions. Therefore it is likely that assertions about the cutting performance can be made based on episodic memories in order to enhance the task performance by adapting the parameters of the controller.

For the long-term, the findings of this work will be utilized to advance the development of cognitive robots. However, before these robots become available for the general public, they will most likely emerge in more distinct domains such as space exploration, healthcare applications, and industrial manufacturing.

The industrial sector has already adopted some of the features of modern robotics. The current trends of the *Internet of Things* and *Industry 4.0* show that the future of manufacturing will be highly concerned with the representation and processing of big amounts of data. The knowledge system developed within this work is capable of handling this demand in a small scale experiment (Nottensteiner et al. 2016). It remains to be seen how semantically annotated episodic memories (e. g. represented as openEASE logs) can be utilized to continuously monitor the task performance of industrial robots by means of haptic effect inference, how the performance can be enhanced, and how failure situations can be detected and predicted, respectively. With this in mind, it will be investigated how the concept of Intelligent Physical Compliance can be adapted for assembly tasks.

The healthcare sector is especially interested in semi-autonomous robotic manipulation w. r. t. the aid of severely impaired persons. People suffering from neuromuscular diseases, stroke, or trauma are often unable to manage their daily life independently and become reliant on 24-h care. In this situation, an assistive robotic manipulator mounted on a wheelchair can provide help and relief. It has been shown that a *Brain-Computer Interface (BCI)* (e. g. surface electromyography electrodes) can be utilized to control such a device, yet it is arduous and difficult (Hochberg et al. 2012; Vogel et al. 2013). To overcome this issue, the reasoning methods developed in this work are currently being adapted to augment the continuous commands of a BCI in order to guide the user in its intended action (Hagengruber et al. 2017). According to the demands of potential users, the most desired tasks for such a robot include physical contact with the person itself, such as eating and drinking, body posture adjustments, and personal hygiene. This close contact to humans makes it mandatory for these robots to react safely at all times, which requires compliant control and intelligent decisions, i. e. Intelligent Physical Compliance.

This work has already contributed to the space sector by providing a novel, context-sensitive UI based on high-level commands within the METERON SUPVIS Justin experiment. Rollin' Justin can be commanded from aboard the ISS to maintain a solar panel farm located on Earth in this fashion. It has been proven that this type of interface is ready to be utilized in future space exploration missions. The necessity for Intelligent Physical Compliance is grounded in the nature of the unknown environment and the delicate instruments. More experiments are scheduled to be conducted in the future to identify the limits of this approach w. r. t. the usability and reliability in case of higher delay and lower bandwidth, the complexity of the task, and the number of simultaneously controllable robots exceeding the maximum cognitive load of an astronaut. The findings may one day contribute to the exploration and colonization of Mars and beyond.

References

Gibson, James J. 1966. The Senses Considered as Perceptual Systems.

Gibson, James J. 1986. *The Ecological Approach to Visual Perception*. Routledge.

Hagengruber, Annette, Daniel Leidner, and Jörn Vogel. 2017. EDAN-EMG-controlled Daily Assistant. In *Proceedings of the ACM/IEEE International Conference on Human-Robot Interaction (HRI)*, 409.

Hochberg, Leigh R., Daniel Bacher, Beata Jarosiewicz, Nicolas Y. Masse, John D. Simeral, Joern Vogel, Sami Haddadin, Jie Liu, Sydney S. Cash, Patrick van der Smagt, et al. 2012. Reach and Grasp by People with Tetraplegia Using a Neurally Controlled Robotic Arm. *Nature* 485 (7398): 372–375.

Lenz, Ian, Ross A. Knepper, and Ashutosh Saxena. 2015. DeepMPC: Learning Deep Latent Features for Model Predictive Control. In *Proceedings of the Robotics: Science and Systems Conference (RSS)*.

Nottensteiner, Korbinian, Tim Bodenmueller, Michael Kassecker, Maximo A. Roa, Andreas Stemmer, Theodoros Stouraitis, Daniel Seidel, and Ulrike Thomas. 2016. A Complete Automated Chain for Flexible Assembly using Recognition, Planning and Sensor-Based Execution. In *Proceedings of the 47st International Symposium on Robotics (ISR)*, 1–8. VDE.

Vogel, Jörn, Justin Bayer, and Patrick Van Der Smagt. 2013. Continuous Robot Control using Surface Electromyography of Atrophic Muscles. In *Proceedings of the IEEE/RSJ International Conference on Intelligent Robots and Systems (IROS)*, 845–850. IEEE.

The manufacturer's authorised representative in the EU is Springer
Nature Customer Service Centre GmbH, Europaplatz 3, 69115 Heidelberg,
Germany. If you have any concerns regarding our products, please
contact ProductSafety@springernature.com

Printed and bound by CPI Group (UK) Ltd, Croydon, CR0 4YY
30/09/2026
02225229-0004